# GÖDEL'S THEOREM IN FOCUS

# PHILOSOPHERS IN FOCUS SERIES

GÖDEL'S THEOREM IN FOCUS
Edited by S.G. Shanker

# Gödel's Theorem in focus

Edited by S.G. Shanker

London and New York

First published in 1988 by
Croom Helm Ltd

Reprinted 1989, 1990, 1991 by
Routledge
11 New Fetter Lane, London EC4P 4EE
29 West 35th Street, New York, NY 10001

British Library Cataloguing in Publication Data

Gödel's theorem in focus.
1. Gödel's theorem
I. Shanker, S.G.
511.3    QA9.65

ISBN 0–7099–3357–6
ISBN 0–415–04575–4 (Pbk)


**Library of Congress Cataloging-in-Publication Data**

ISBN 0–7099–3357–6
ISBN 0–415–04575–4 (Pbk)


Typeset in 10pt Times Roman by Leaper & Gard Ltd, Bristol
Printed and bound in Great Britain by
Mackays of Chatham PLC, Chatham, Kent

# Contents

# Preface

An ever-growing number of sub-disciplines in philosophy — ranging from the philosophies of science and language to the philosophy of mind and aesthetics — now demand a working acquaintance with Kurt Gödel's incompleteness theorems from their students. For Gödel's theorems raise issues which lie at the very heart of modern attempts to revitalise metaphysics and/or the Mechanist Thesis. Unfortunately, given the highly technical nature of Gödel's proof these debates have remained relatively inaccessible to those not trained in mathematical logic. The present book has been designed to meet these needs by providing a lucid introduction to the mechanics and mathematical import of Gödel's proof. We begin with a short biographical sketch of Kurt Gödel by John W. Dawson, Jr., followed by Stephen Kleene's overview of Gödel's work in mathematical logic. With this background in place we then address the mounting controversy in the philosophy of mathematics surrounding the philosophical significance of Gödel's theorems.

Some will no doubt regard the latter phenomenon as a reflection of the inevitable time-lag between scientific and mathematical discoveries versus philosophical comprehension. Perhaps it will even be seen to corroborate the increasingly popular thesis in the sociology of knowledge that it is only once a science has digested the full implications of a breakthrough that it becomes the property of philosophers to exaggerate and distort. But unlike standard mathematical results Gödel's theorems are inextricably linked to the epistemological disputes which they have sparked off; indeed, nowhere could this be more evident than in the writings of Gödel himself. It is not surprising, therefore, that Gödel's theorems should present us with a catalogue of philosophical problems, many of which we are only just beginning to recognise, let alone resolve.

To begin with there is the anomalous reception of Gödel's first incompleteness theorem; in the words of John W. Dawson, Jr., how was it that 'one of the most profound discoveries in the history of logic and mathematics was assimilated promptly and almost without objection by Gödel's contemporaries'? As Solomon Feferman shows us, this issue is intimately connected with the larger question of how Gödel's subsequent work in mathematical logic and the growing confidence with which he expressed his platonist convictions relate to his earlier interpretation and presentation of the second incompleteness theorem. This brings us to the point where we can address the main concerns of the final three papers. First Michael Resnik outlines the seriousness of the sceptical problem created by Gödel's theorems. Michael Detlefsen then takes up the challenge of attempting to rescue Hilbert's Programme from the impasse in which Gödel apparently left it. In the final paper I examine the nature of this dilemma in the light of Wittgenstein's attempt to resolve by dissolving the crisis created by Gödel's theorems.

Like all philosophical sceptical problems the issues raised by Gödel's theorems are pregnant with possibilities and fraught with dangers. Chief amongst the latter is an inevitable tendency to become distanced from the *fons et origo* of these developments. For it becomes ever more tempting and acceptable to rely on the findings of commentators who might themselves have based their readings on earlier summaries. To be sure, it is common practice to accept the verdict obtained by the experts in a field without inspecting their findings. But such custom presupposes concord. The ultimate aim of this book has been to outline the basis of the consensus which has hitherto obtained in order then to question it. It is our hope that this will stimulate renewed interest in the ongoing interpretation of the significance of Gödel's incompleteness theorems.

S.G. Shanker
Atkinson College, York University

# Acknowledgements

I am deeply indebted to John W. Dawson, Jr. and Michael Detlefsen for their generous assistance in the design of this book. I have also benefited greatly from many fruitful discussions with Rainer Born. Lastly I would like to thank the Canada Council for the Postdoctoral Fellowship which enabled me to work on this project.

We are grateful to the authors and their publishers for permission to reprint the following articles:

John W. Dawson, Jr., 'Kurt Gödel in Sharper Focus', *The Mathematical Intelligencer*, vol. 6, no. 4 (1984), by permission of Springer-Verlag.

John W. Dawson, Jr., 'The Reception of Gödel's Incompleteness Theorems', *Philosophy of Science Association*, vol. 2 (1984). Copyright © 1985 by the Philosophy of Science Association.

Michael Detlefsen, 'On Interpreting Gödel's Second Theorem', *Journal of Philosophical Logic*, vol. 8, no. 3 (June 1979), pp. 297-313. Copyright © 1979 by D. Reidel Publishing Company, Dordrecht, Holland.

Solomon Feferman, 'Kurt Gödel: Conviction and Caution', a revised and expanded version of the author's contribution to Section 13 of the 7th International Congress of Logic, Methodology and Philosophy of Science (held in Salzburg, 11-16 July 1983), and published in a separate issue of *Philosophia Naturalis*, Paul Weingartner and Christine Pühringer (eds.), *Philosophy of Science — History of Science. A Selection of Contributed Papers of the 7th International Congress of Logic, Methodology and Philosophy of Science, Salzburg, 1983*, Meisenheim/Glan, Verlag Anton Hain (1984).

Kurt Gödel, 'On Formally Undecidable Propositions of *Principia Mathematica* and Related Systems I', Jean van Heijenoort (trans.), in Jean van Heijenoort (ed.), *From Frege to Gödel: A Source Book in Mathematical Logic, 1879-1931* (Cambridge, Mass., Harvard University Press, 1967). Copyright © 1967 by the President and Fellows of Harvard College.

Stephen C. Kleene, 'The Work of Kurt Gödel', *The Journal of Symbolic Logic*, vol. 41, no. 4 (1976).

Stephen C. Kleene, 'An Addendum to "The Work of Kurt Gödel"', *The Journal of Symbolic Logic*, vol. 41, no. 4 (1978).

Michael D. Resnik, 'On the Philosophical Significance of Consistency Proofs', *Journal of Philosophical Logic*, vol. 3/1-2 (1974), pp. 133-47. Copyright © 1974 by D. Reidel Publishing Company, Dordrecht, Holland.

# I

# Kurt Gödel in Sharper Focus

John W. Dawson, Jr.

## 1. INTRODUCTION

The lives of great thinkers are sometimes overshadowed by their achievements — a phenomenon perhaps no better exemplified than by the life and work of Kurt Gödel, a reclusive genius whose incompleteness theorems and set-theoretic consistency proofs are among the most celebrated results of twentieth-century mathematics, yet whose life history has until recently remained almost unknown.

Several tributes to Gödel have appeared since his death in 1978, most notably the obituary memoirs of Curt Christian, Georg Kreisel, and Hao Wang (Christian, 1980; Kreisel, 1980; Wang, 1978).[1] But none of those authors developed a close personal acquaintance with Gödel before the 1950s, and there are discrepancies among their accounts. To resolve them, to substantiate or refute various rumors that have circulated, and to learn further details about Gödel's life and work, scholars must therefore turn to primary documentary sources. In the remainder of this article I shall highlight some aspects of Gödel's life that have been thrown into sharper focus as the result of my own explorations among such sources; I have drawn primarily on my experiences over the past two years in cataloguing Gödel's *Nachlass* at the Institute for Advanced Study in Princeton, supplemented by personal interviews, visits, and correspondence with various individuals here and abroad. Of necessity, I have repeated some biographical information already available in the memoirs cited. My aim, however, has been to amplify or correct details of those accounts, insofar as primary documentation is available.

I am grateful to the Institute for Advanced Study for permission to quote from, and to reproduce photographs of, unpublished items in the *Nachlass*; to Rudolf Gödel, Kurt's brother and only surviving close relative, for his gracious responses to my inquiries; and to H. Landshoff for assistance in preparing illustrations for this article.

## 2. THE GÖDEL *NACHLASS*: PROVENANCE, ARRANGEMENT, AND DISPOSITION

The scientific *Nachlass* of Kurt Gödel, including correspondence, drafts, notebooks, unpublished manuscripts, books from his library, and all manner of loose notes and memoranda was donated to the Institute for Advanced Study after Gödel's death by his widow. Gödel himself made no provision for the disposition of his papers, although correspondence in the *Nachlass* indicates that the Library of Congress solicited them from him. Indeed, his attitude toward posthumous preservation or publication of his papers seems to have been ambivalent. Thus he spoke to Dana Scott of his desire to have certain papers published posthumously and even asked Scott to prepare typescripts of some of them. Yet on the other hand, he declined several invitations to consider publication of his collected works, maintaining that the most important of them were readily available and that the rest were only of historical and biographical interest. Ultimately, his papers were gathered into boxes and placed in a cage in the basement of the Institute's historical studies library (which has no archival facilities) to await further disposition.

My own involvement with Gödel's papers began in the fall of 1980. In an effort to track down Gödel's lesser-known publications, I consulted the bibliography in Bulloff, Holyoke and Hahn (eds, 1969), prepared on the occasion of Gödel's sixtieth birthday. Though I presumed that that listing would be complete, a colleague soon called my attention to an item not cited there, and I later found two more. Subsequently, I wrote to the IAS to inquire whether Gödel himself had ever prepared a bibliography of his own work; in response, I received a typewritten list identical to that in Bulloff, Holyoke and Hahn. Concluding that no comprehensive bibliography had ever been attempted, I undertook to compile one myself, at the same time embarking on the more ambitious program of translating all of Gödel's previously untranslated works into English.[2] Eventually I was offered the opportunity of cataloguing the *Nachlass*, and in June 1982 I arrived in Princeton to begin work.

At the outset, there were three major problems to be faced: the sheer bulk of the materials comprising the *Nachlass*; my lack of training in archival technique; and the necessity of penetrating Gödel's Gabelsberger shorthand, an obsolete German script he used extensively.[3] It was at first somewhat daunting to find that the *Nachlass* occupied two large filing cabinets plus some *sixty*-odd cardboard packing boxes. The majority of the boxes, however, contained books from Gödel's library, back issues of journals he received, and reprints and offprints sent to him by others. The remaining, much more manageable, body of primary documentary material required about three months to survey in order to devise an appropriate arrangement scheme. (See Dawson (1983b) for further details.)

My ignorance of archival principles was remedied by consultations with archivists and by reference to Gracy's helpful manual (1977). Gödel's eminently logical mind, extremely methodical habits, and clear handwriting were also a boon to my efforts. In particular, one fundamental archival conflict — that of preserving the original order of a manuscript collection while facilitating scholarly access to it — has seldom arisen. My task has rather been that of restoring Gödel's order to papers gathered somewhat haphazardly after his death.

The shorthand problem proved the most difficult. Early on, my wife volunteered to learn the script, provided suitable instruction manuals could be found. At the same time we began a search for 'native' stenographers. Eventually, both approaches proved successful. In the *Nachlass* itself, Gödel's own shorthand textbook turned up, along with several 'Rosetta stones' — vocabulary notebooks in several languages with foreign words in longhand paired with their German synonyms in Gabelsberger — that allowed comparison of Gödel's individual 'hand' with the textbook examples. Much later, we also located a German émigré, Hermann Landshoff, who had learned the Gabelsberger script as a youth and was willing to assist us.

Cataloguing of the *Nachlass* should be completed by the summer of 1984. After that, it is expected that the papers will be donated to the library of Princeton University, where they will be made available to scholars.

## 3. GÖDEL'S CHILDHOOD AND YOUTH

Kurt Friedrich Gödel was born 28 April 1906 in Brünn, Moravia (then part of the Austro-Hungarian empire; now Brno, Czechoslovakia), the second of the two children of Rudolf and Marianne Gödel. Then as now, Brünn was a major textile centre, and Kurt's father worked as director of the Friedrich Redlich textile factory. (Redlich himself was Kurt's godfather, later killed by the Nazis. From him the infant's middle name was presumably taken. Gödel officially dropped his middle name when he became a U.S. citizen, but the initial 'F.' survives on his tombstone.) Patent correspondence preserved among Gödel's papers attests to his father's inventiveness in the textile field, and even Marianne Gödel's maiden name, Handschuh ('glove'), suggests the garment trade. (As noted in Christian (1980) and Kreisel (1980), her father was in fact a weaver.)

The bare facts about Gödel's birth are recorded on a copy of his *Taufschein* (baptismal certificate), preserved in the *Nachlass* along with his naturalization papers. That document shows that he was born at 5 Bäckergasse (now Pekařska) and baptized in the German Lutheran congregation of Brünn. Later the family moved to a villa at 8A Spilberggasse (now Pellikova), a residence more befitting their moderately wealthy circumstances.

Gödel's ethnic heritage was thus neither Czech nor Jewish, as has sometimes been asserted. Though his parents were both born in Brünn, they were part of the German community there, and the children attended German-language schools. Kurt did not enrol in optional courses in the Czech language, and both he and his brother gave up their postwar Czech citizenship after they became students at the University of Vienna.

Other items in the *Nachlass* pertaining to Gödel's youth include report cards from both elementary and secondary schools and several of his school notebooks. Particularly quaint is his first arithmetic workbook, which contains but a single error in computation. The report cards are indicative of curricula of the time, which laid heavy stress on science and languages, in addition to such required courses as religion, drawing, and penmanship. Latin and French were required, and Gödel chose English as the second of his two elective subjects (after shorthand). In general, he seems to have been quite interested in languages. His library contains many foreign-language dictionaries, and there are vocabulary and exercise notebooks in Italian and Dutch in addition to the languages already mentioned.

All of Gödel's school records attest to his diligence and outstanding performance as a student. Indeed, only once did he receive less than the highest mark — in mathematics. But the report cards also record a rather large number of excused absences, including exemptions during the years 1915-16 and 1917-18 from participation in physical education. The earlier of those exemptions probably corresponds to a childhood bout with rheumatic fever, an episode Rudolf Gödel believes to have been the source of his brother's later hypochondria.

Some later material in the *Nachlass* also relates to Gödel's youth. One especially valuable item is a questionnaire sent to Gödel in 1974 by the sociologist Burke D. Grandjean — a document Gödel dutifully filled out but never returned. In response to some of its queries, Gödel noted that his interest in mathematics began at about age 14, stimulated by an introductory calculus text in the well-known Göschen collection; that his family was little affected by World War I and the subsequent inflation; that he was never a member of any religious congregation, although he was a believer (describing himself as a theist rather than pantheist, 'following Leibniz rather than Spinoza'); and that prior to his enrolment at the University of Vienna he had little contact with Vienna's intellectual or cultural life except through the newspaper *Neue Freie Presse*.

## 4. VIENNA YEARS, AND VISITS TO PRINCETON

Aside from his doctoral diploma and some of his course notebooks, the *Nachlass* contains few records of Gödel's university career. According to his own account (again in response to the Grandjean questionnaire), he entered the University of Vienna in 1924 intending to major in physics. Once there, however, he was influenced by the mathematical lectures of Philipp Furtwängler and the lectures of Heinrich Gomperz on the history of philosophy. He switched into mathematics in 1926. At about the same time, under the guidance of Hans Hahn, he began to attend meetings of the Vienna Circle, with whose views, however, he disagreed, and from which he later took pains to dissociate himself (as revealed, for example, in several letters found in the *Nachlass*). Gödel submitted his dissertation in the autumn of 1929, a year marked not only by worldwide financial collapse, but by the premature death of Gödel's father on 23 February, five days before his fifty-fifth birthday. On 6 February 1930, Gödel's doctorate in mathematics was conferred by the University of Vienna (not, as E.T. Bell asserts in his *Mathematics, Queen and Servant of Science*, by 'the University of Brno in engineering').

Shortly afterward, in an attempt to pursue the aims of Hilbert's Programme, Gödel sought to find an interpretation of analysis within arithmetic. In so doing, he came to realize that the concept of provability could be defined arithmetically. This led to his incompleteness proof, which, ironically, overturned Hilbert's Programme (at least as originally envisioned). Yet Gödel announced his momentous discovery almost casually, toward the end of a dicussion on foundations at a conference in Königsberg where only the day before he had lectured on his dissertation results (the completeness of the first-order predicate calculus).[4] Reaction (not always accompanied by understanding) was immediate, ranging from profound appreciation by von Neumann (who two months later nearly anticipated Gödel's discovery of the *second* incompleteness theorem) to vigorous criticism by Zermelo (see Dawson (reprint) and Grattan-Guinness (1979)) and even to a claim to priority by Finsler (see van Heijenoort's note, pp. 438-440 in van Heijenoort (ed., 1967)), dismissed by Gödel with uncharacteristic disdain. The incompleteness paper appeared early in 1931. Later Gödel submitted it to the University of Vienna as his *Habilitationsschrift*, thereby earning the right to teach as *Privatdozent*. In the meantime he participated actively in Karl Menger's colloquium, where he presented nearly a dozen papers and collaborated in editing volumes 2-5 and 7-8 of the colloquium proceedings (*Ergebnisse eines mathematischen Kolloquiums*).

Officially, Gödel's tenure as *Privatdozent* extended from 1933 to 1938. In fact, however, his lecturing at the University of Vienna was repeatedly interrupted, both by visits to America and by episodes of ill health. Indeed, on the basis of enrolment slips saved by Gödel and records of the University of Vienna, it appears that Gödel actually taught only three courses there: foundations of arithmetic, in the summer of 1933; selected chapters in mathematical logic, in the summer of 1935; and axiomatic set theory, in the spring of 1937.

From the published memoirs it is difficult to piece together a coherent chronology of Gödel's visits to America; there were actually three such prior to his emigration in 1940. He first came in 1933-34 to lecture on his incompleteness theorems at the Institute for Advanced Study, where he spent the academic year. It was the Institute's first year of operation, without a building of its own and with titles for the visiting scholars yet to be decided upon. The official IAS *Bulletin* for that year lists Gödel simply as a 'worker'. In April, Gödel also travelled to New York and Washington, where he lectured before the Philosophical Society of New York and the Washington Academy of Sciences.

After his return to Europe Gödel suffered a nervous breakdown. He entered a sanatorium and was forced to postpone an invitation to return to the IAS for the second term of 1934-35. In the meantime he began his investigations in set theory; and when he did return to the IAS in October 1935, he told von Neumann of his consistency proof for the axiom of choice. A month later Gödel suddenly resigned, suffering from depression and overwork. Veblen saw him aboard ship in New York and telegraphed ahead to Gödel's family. More time in sanatoria followed, and, as noted above, Gödel only resumed his teaching in Vienna in the spring of 1937.

Later that summer,[5] Gödel saw how to extend his consistency proof to the generalized continuum hypothesis. In the fall of 1938 he returned once more to America, spending the first term at the IAS and the second, at Menger's invitation, at Notre Dame. (Not 'Rotterdam', as stated in Wang (1981)). At both institutions he lectured on his consistency results, and at Notre Dame he and Menger offered a joint course on elementary logic. The *Nachlass* contains manuscripts of all those lectures — carefully written in English, except for a single page of examination questions, effectively concealed in Gabelsberger.

## 5. EMIGRATION AND AMERICAN CAREER

Gödel intended to return again to Princeton in the fall of 1939, but personal and political events intervened. The previous September, only about two weeks before his departure for America, he had married Adele Nimbursky (nee Porkert) in

Vienna. Though he had known Adele for over a decade, their marriage had been delayed by opposition from Gödel's family; for she was not only a divorcee, older than Kurt, but she had worked as a dancer and was somewhat disfigured by a facial birthmark. It proved nevertheless to be an enduring union. Their first year of marriage, however, was spent apart, as Adele remained behind in Vienna during the academic year 1938-39.

After his return to Vienna to rejoin his bride in the summer of 1939, Gödel was called up for a military physical by the Nazi government. Writing to Veblen in November, Gödel reported that contrary to his expectation, 'I was mustered and found fit for garrison duty.'[6] At the same time, to retain his right to teach at the University of Vienna, he was obliged to apply to Nazi authorities for appointment as a *Dozent neuer Ordnung,* thereby subjecting himself to political and racial scrutiny; and though his mother and brother lived unmolested in Brno and Vienna throughout the Nazi occupation, Gödel was suspect because of his association with Jewish intellectuals such as Hahn. Eventually his application was approved, but only after his emigration to America.

Even there, however, he was thought by many to be Jewish. Thus Bertrand Russell declared in the second volume of his *Autobiography*:

> I used to go to [Einstein's] house, once a week to discuss with him and Gödel and Pauli. These discussions were in some ways disappointing, for, although all three of them were Jews and exiles and, in intention, cosmopolitans, I found that they all had a German bias toward metaphysics ... Gödel turned out to be an unadulterated Platonist, and apparently believed that an eternal 'not' was laid up in heaven, where virtuous logicians might hope to meet it hereafter.

In 1971, Gödel's attention was called to this passage by Kenneth Blackwell, curator of the Russian archives at McMaster University. Gödel drafted a reply (never actually sent) that is preserved in the *Nachlass*:

> As far as the passage about me [in Russell's autobiography] is concerned, I have to say *first* (for the sake of truth) that I

am not a Jew (even though I don't think this question is of any importance), 2.) that the passage gives the wrong impression that I had many discussions with Russell, which was by no means the case (I remember only one). 3.) Concerning my 'unadulterated' Platonism, it is no more 'unadulterated' than Russell's own in 1921 when in the *Introduction [to Mathematical Philosophy]* he said '[Logic is concerned with the real world just as truly as zoology, though with its more abstract and general features].' At that time evidently Russell had met the 'not' even in this world, but later on under the influence of Wittgenstein he chose to overlook it.

(In Gödel's draft, only an ellipsis (...) appears between the quotation marks. The bracketed passage inserted here was quoted by Gödel in his 1944 essay 'Russell's mathematical logic'.)

Somehow, in the midst of the political turmoil, Gödel succeeded in obtaining exit visas: his passport, preserved in the *Nachlass*, testifies to his frantic efforts to obtain transit documents from consulates in Vienna and Berlin. By then it was too dangerous to risk crossing the Atlantic, so, in January 1940, he and Adele travelled through Lithuania and Latvia to board the trans-Siberian railway at Bigosovo. After crossing Russia and Manchuria they made their way to Yokohama and thence by ship to San Francisco, where they arrived 4 March 1940.

At the IAS Gödel found an intellectual haven from which he was rarely to venture again. Reclusive by nature, he seems not to have minded (and perhaps even to have sought) his growing isolation. Several of Gödel's acquaintances, however, have remarked that Princeton society proved unreceptive (not to say hostile) toward Adele, and she appears to have led a very lonely life there.

Professionally, the IAS offered Gödel relative job security. Yet he was appointed on a yearly basis until 1946, when he was finally made a permanent member. Only in 1953, five years after becoming a U.S. citizen and two years after sharing the first Einstein award,[7] was he promoted to professor. Gödel himself seems never to have expressed dissatisfaction about this long delay, but others have called for an explanation. In particular, Stanislaw Ulam (1976, p. 80) and Freeman Dyson (1983, p. 48) have brought the matter to public attention. Ulam, indeed, quoting remarks made to him by von Neumann, has suggested that Gödel's treatment was occasioned by the personal opposition of some unnamed IAS colleague. The *Nachlass* itself sheds no light on the matter, but some 'old-timers' at the Institute have suggested that a division of opinion prevailed among Gödel's colleagues: some felt that Gödel would not welcome the administrative responsibilities entailed by faculty status, while others feared that if he were promoted, his sense of duty and legalistic habit of mind might impel him to undertake such responsibilities all too seriously, perhaps hindering efficient decision-making by the faculty. In the event, such fears seem to have been justified; but one should also note Gödel's own statement in 1946 in a letter to C.A. Baylis, in which he noted that 'apprehension that cooperation of this kind [service in offices or on committees] would be expected' was 'the very reason' he had so belatedly joined the Association for Symbolic Logic.

In any case, the Institute allowed Gödel freedom to pursue a broad range of intellectual interests. At first he labored to prove the independence of the axiom of choice and the continuum hypothesis, but the latter especially proved unyielding, and eventually he gave up the attempt.[8] Instead, he turned to philosophy. The transition is marked by his 1944 essay 'Russell's mathematical logic', solicited by P.A. Schilpp for his *Library of Living Philosophers* series. Subsequently, Schilpp was to solicit essays for the Einstein, Carnap, and Popper volumes of the series as well. Gödel accepted all but the last, and he devoted great care to each — so much so that in every case his essay was among the last to be received. Schilpp displayed great patience and diplomacy, but Gödel could not be hurried; and when his Russell essay was received too late for Russell to reply, Gödel considered withdrawing it altogether. Eventually he yielded to Schilpp's entreaties, but when the situation recurred a few years later Schilpp was forced to send the Carnap volume to press without Gödel's contribution. It remains unpublished in the *Nachlass*.

In contrast, Gödel's contribution to the Einstein volume was not only commented upon, but praised by Einstein as marking a significant advance in the physical, as well as philosophical, understanding of relativity; for Gödel had in fact discovered an unexpected solution to Einstein's field equations of gravitation, one permitting 'time travel' into the past.[9] The contribution to Schilpp's anthology is quite brief, but Gödel prepared a much longer essay that has remained unpublished. It too is preserved in the *Nachlass* — in six different versions. It is worth noting that Gödel's interest in relativity went beyond the purely theoretical. In his essay he argues in favour of the possible relevance of his models to our own world, and in the *Nachlass* there are two notebooks devoted to tabulations of the angular orientations of galaxies (which Gödel hoped might exhibit a preferred direction). Freeman Dyson has remarked that even much later, Gödel maintained a keen interest in such observational data.

Yet another unpublished item in the *Nachlass* is the text of Gödel's Gibbs Lecture to the American Mathematical Society, delivered at Brown University on 26 December 1951. Titled 'Some basic theorems on the foundations of mathematics and their philosophical implications', it is Gödel's contribution to the debate on mechanism in the philosophy of mind.

In addition to such relatively finished papers, there are a great many pages of notes by Gödel, including 16 mathematical workbooks, 14 philosophical notebooks, and voluminous shorthand notes on Leibniz. The latter appear to be partly bibliographic, but a recently discovered memorandum suggests that there are about 1000 pages of Gödel's own philosophical assertions as well.

Gödel's long-standing interest in Leibniz is also indicated by an extensive correspondence with archivists, conducted jointly with Oskar Morgenstern during the years 1949-53. The object of the correspondence was the microfilming of some of Leibniz's unpublished manuscripts in Hannover, with the aim not only of preserving them, but of making them available to American scholars. Ultimately the attempt failed, but copies of the manuscripts were later deposited at the University of Pennsylvania through the independent efforts of Prof. Paul Schrecker.

## 6. LATER YEARS

Gödel's last published paper appeared in 1958. Based on results obtained nearly eighteen years earlier (on which Gödel lectured at Yale, 15 April 1941), it presented a consistency proof for arithmetic by means of 'a [t]hitherto unused extension' of principles formulated in intuitionistic mathematics. As such, it represented a return to earlier mathematical interests (and to the German language — it was the only paper Gödel published in German after his emigration); nevertheless, it is decidedly philosophical in character and was published in the philosophical journal *Dialectica.* Unfortunately, the paper is notoriously difficult to translate. In the early 1970s Gödel himself prepared a revised and expanded English version that reached the stage of galley proofs but never appeared.

After 1958, Gödel devoted himself to revisions of earlier papers, to a search for new axioms to settle the continuum hypothesis (in the wake of Cohen's independence proofs), and to a study of the philosophy of Edmund Husserl. Honorary degrees, academy memberships, and awards (including the National Medal of Science in 1975) were bestowed upon him from many quarters, but he became increasingly withdrawn and preoccupied with his health. He consulted doctors but distrusted their advice, which he often failed to heed. Thus, in the late 1960s, he refused recommended surgery for a prostate condition, despite the urgings of concerned colleagues; and earlier, in the 1940s, he delayed treatment of a bleeding duodenal ulcer until life-saving blood transfusions had to be administered. Afterwards he observed a strict dietary regimen, and as the years wore on his figure became even more gaunt.

During the final decade of Gödel's life, his wife underwent surgery and suffered two strokes that led to her placement in a convalescent home. By all accounts, Gödel attended her devotedly, but he soon began to display signs of depression and paranoia. Correspondence, even with is brother, virtually ceased during the last two years of Gödel's life. In the end, his paranoia conformed to a classic syndrome: fear of poisoning leading to self-starvation. After a relatively brief hospitalization, Gödel died 14 January 1978, of 'malnutrition and inanition' caused by 'personality disturbance'. (His death certificate, from which these phrases are quoted, is on file in the Mercer County courthouse, Trenton, N.J.). He is buried beside his wife and mother-in-law in the historic Princeton Cemetery.

## 7. PROSPECTS

To what extent can study of Gödel's *Nachlass* be expected to reveal hitherto unknown discoveries? It seems unlikely that major new mathematical results will be found in Gödel's notebooks, although the reputed 'general' consistency proof for the axiom of choice mentioned in Wang (1978) and Wang (1981) may prove to be an exception — if, indeed, it can be reconstructed. Gödel was certainly cautious and overly fastidious in submitting his work for publication, but there is no evidence that he actively withheld important mathematical discoveries; and though details of his researches remain largely concealed behind his shorthand, the topics of his investigations can nonetheless be determined to a large extent from (longhand) headings in the notebooks. On that basis, it seems safe to predict that some lesser results of mathematical interest will be found there, along with some anticipations of or alternative approaches to results of others (such as Gödel's early recognition of errors in Herbrand's work, cited in Kreisel (1980), or his partial independence results in set theory). Of course, details of Gödel's explorations should be of great interest to mathematical historians. Beyond that, I would venture to predict that of all the unpublished material in the *Nachlass*, Gödel's philosophical investigations will turn out to be of greatest interest. Certainly they figure most prominently among the items he left in relatively finished form and that he himself considered to be potentially publishable.

Plans for publication of Gödel's collected works are now well under way.[10] Two volumes are presently envisioned, under the editorship of Solomon Feferman (editor-in-chief), myself, Stephen C. Kleene, Gregory H. Moore, Robert M. Solovay, and Jean van Heijenoort. The first volume, now in preparation, will contain all of Gödel's published articles and reviews, together with his doctoral dissertation (in its original unpublished form) and his revised English version of the *Dialectica* paper, as well as three short notes appended to galley proofs of the latter. Papers in German will be accompanied by English translations on facing pages, and each article will be preceded by introductory commentary. Textual notes, a short biographical essay, and an extensive bibliography will round out the volume.

Detailed contents of Volume 2 remain to be determined, subject both to the success of our decipherment efforts and to our ability to obtain necessary funding and copyright permissions. We hope, however, to include all the relatively finished papers mentioned previously, plus other lecture texts, excerpts from the mathematical notebooks, and selected correspondence, including not only extensive exchanges with other mathematicians, but individual letters of interest. Should there

13

be enough material to warrant it, further volumes may also be considered. The editors will welcome contributions of correspondence with Gödel or recollections of him.

## NOTES

1. Wang's later memoir (Wang, 1981) contains some interesting additional information, but on the whole it is less reliable, even though it alone was written before Gödel's death and was submitted to Gödel for his approval and correction. In particular, it is marred by several errors in references to dates and places.

2. The bibliography has now appeared (Dawson, 1983a); note also Dawson (1984a). The translations will be included in the first volume of our edition of Gödel's collected works.

3. Devised by Franz Xaver Gabelsberger, for whom it is named, the script was one of two competing German shorthand systems in widespread use during the early decades of this century. Eventually the two systems merged to form the modern *Einheitskurzschrift* ('unified shorthand'); alas, however, those trained only in the modern system can read *neither* the Gabelsberger nor Stolze–Schrey scripts from which it was derived. Yet there is a need for some younger scholars to learn to read these scripts, since many prominent intellectuals used them in their daily lives — not, as is often supposed, for reasons of secrecy, but as efficient means for rapid and concise recording of events and ideas.

4. An abridged transcript of the discussion appeared in *Erkenntnis* 2 (1931), pp. 135-151; for an English translation and commentary, see Dawson (1984b).

5. This date is based on Gödel's correspondence with von Neumann. On 13 July 1937, von Neumann wrote Gödel from Budapest, saying that he expected to visit Vienna in a few weeks, and that while there he hoped to speak with Gödel and learn more about his plans. In the same letter he urged Gödel to consider publishing his work on the axiom of choice in the *Annals of Mathematics*. In his next letter, however, written 14 September from New York, von Neumann advised Gödel that the editors of the *Annals* were prepared to expedite publication of his work on the generalized continuum hypothesis. In the end, both consistency results were announced late the following year in the *Proceedings of the National Academy of Sciences*.

6. By 'mustered' Gödel apparently meant only that he had to report for the physical examination. It seems very unlikely that he was actually sworn in.

7. With Julian Schwinger; not, as stated in Quine (1978), with von Neumann. Von Neumann was actually a member of the awards committee, and he may have introduced Gödel's name for consideration. In any case, there is evidence in von Neumann's papers that Schwinger alone was originally proposed for the award.

8. Rumours have persisted that Gödel actually obtained the independence of the axiom of choice in the early 1940s but refused to

publish his results. In particular, after Cohen's proofs in 1964, Mostowski asserted, 'Es ist seit 1938 bekannt, dass Gödel einen Unabhängigkeitsbeweis dieser Hypothesen besitzt; trotz vielen Anfragen verriet aber nie sein Geheimnis' (*Elemente der Mathematik* 19, p. 124). But Gödel himself denied this. In a letter to Wolfgang Rautenberg (published in *Mathematik in der Schule 6*, p. 20) he stated explicitly:

> Die Mostowskische Behauptung ist insofern unrichtig, als ich bloss in Besitzte gewissen Teilresultate war, nämlich von Beweisen für die Unabhängigkeit der Konstruktibilitäts- und Auswahlaxioms in der Typentheorie. Auf Grund memer höchst unvollständigen Aufzeichnungen von damals (d.h. 1942) könnte ich ohne Schwierigkeiten nur den ersten dieser beiden Beweise rekonstruieren. Meine Methode hat eine sehr nahe Verwandtschaft mit der neuerdings von Dana Scott entwickelten, weniger mit den Cohenschen.

It is clear from Gödel's correspondence that he had great respect for Cohen's work; indeed, he described Cohen's achievements as 'the greatest advance in abstract set theory since its foundation by Georg Cantor.' Nevertheless, the method by which Gödel obtained his partial results may still prove to be of interest.

9. Technical details were published that same year (1949) in *Reviews of Modern Physics*, and a year later Gödel spoke on his results at the International Congress of Mathematicians in Cambridge, Massachusetts.

10. A nearly complete edition of Gödel's published works, excluding reviews, has already appeared in Spanish translation; see Jesús Mosterín, ed., *Kurt Gödel, Obras Completas*, Alianza Editorial, Madrid, 1981.

# REFERENCES

Bulloff, Jack, T.C. Holyoke and S.W. Hahn (eds.) (1969) *Foundations of Mathematics: Symposium Papers Commemorating the Sixtieth Birthday of Kurt Gödel*, New York: Springer-Verlag.

Christian, Curt (1980) Leben und Wirken Kurt Gödels. *Monatshefte für Mathematik* 89, 261-73.

Dawson, J.W., Jr. (1983a) The published work of Kurt Gödel: an annotated bibliography. *Notre Dame Journal of Formal Logic* 24, 255-84.

Dawson, J.W., Jr. (1983b) Cataloguing the Gödel *Nachlass* at the Institute for Advanced Study. *Abstracts of the 7th International Congress of Logic, Methodology, and Philosophy of Science* 6, 59-61.

Dawson, J.W., Jr. (1984a) Addenda and corrigenda to 'The published work of Kurt Gödel'. *Notre Dame Journal of Formal Logic* 25, to appear.

Dawson, J.W., Jr. (1984b) Discussion on the foundation of mathematics. *History and Philosophy of Logic* 5, 111-29.

Dawson, J.W., Jr. (reprint) Completing the Gödel-Zermelo correspondence. *Historia Mathematica*, 12, 66-70.

Dyson, Freeman (1983) Unfashionable pursuits. *The Mathematical Intelligencer* 5:3, 47-54

Feferman, Soloman, John W. Dawson, Jr., Stephen C. Kleene, Gregory H. Moore, Robert M. Solovay and Jean van Heijenoort (eds.) (1986) *Kurt Gödel: Collected Works. Volume I: Publications, 1929-1936*, Oxford: Oxford University Press.

Gracy, David B., II(1977) *Archives & Manuscripts: Arrangement & Description* (Basic Manual Series). Chicago: Society of American Archivists.

Grattan-Guinness, Ivor (1979) In memoriam Kurt Gödel: his 1931 correspondence with Zermelo on his incompletability theorem. *Historia Mathematica* 6, 294-304.

Kreisel, Georg (1980) Kurt Gödel, 1906-1978, elected For. Mem. R.S. 1968. *Biographical Memoirs of Fellows of the Royal Society* 26, 148-224; (1981) corrigenda, 27, 697; (1982) further corrigenda, 28, 718.

Quine, Willard V. (1978) Kurt Gödel (1906-1978). *Year Book of the American Philosphical Society*, 81-84.

Ulam, Stanislaw (1976) *Adventures of a Mathematician*. New York: Scribner's.

van Heijenoort, Jean (ed.) (1967) *From Frege to Gödel, A Source Book in Mathematical Logic, 1879-1931*. Cambridge: Harvard.

Wang, Hao (1978) Kurt Gödel's intellectual development. *The Mathematical Intelligencer* 1, 182-4.

Wang, Hao (1981) Some facts about Kurt Gödel. *The Journal of Symbolic Logic* 46, 653-9.

# II

# On Formally Undecidable Propositions of *Principia Mathematica* and Related Systems I[1] (1931)

## Kurt Gödel

## 1.

The development of mathematics toward greater precision has led, as is well known, to the formalization of large tracts of it, so that one can prove any theorem using nothing but a few mechanical rules. The most comprehensive formal systems that have been set up hitherto are the system of *Principia mathematica* (*PM*)[2] on the one hand and the Zermelo–Fraenkel axiom system of set theory (further developed by J. von Neumann)[3] on the other. These two systems are so comprehensive that in them all methods of proof today used in mathematics are formalized, that is, reduced to a few axioms and rules of inference. One might therefore conjecture that these axioms and rules of inference are sufficient to decide *any* mathematical question that can at all be formally expressed in these systems. It will be shown below that this is not the case, that on the contrary there are in the two systems mentioned relatively simple problems in the theory of integers[4] that cannot be decided on the basis of the axioms. This situation is not in any way due to the special nature of the systems that have been set up but holds for a wide class of formal systems; among these, in particular, are all systems that result from the two just mentioned through the addition of a finite number of axioms,[5] provided no false propositions of the kind specified in note 4 become provable owing to the added axioms.

Before going into details, we shall first sketch the main idea of the proof, of course without any claim to complete precision. The formulas of a formal system (we restrict ourselves here to the system *PM*) in outward appearance are finite sequences of

17

primitive signs (variables, logical constants, and parentheses or punctuation dots), and it is easy to state with complete precision *which* sequences of primitive signs are meaningful formulas and which are not.[6] Similarly, proofs, from a formal point of view, are nothing but finite sequences of formulas (with certain speci-fiable properties). Of course, for metamathematical consider-ations it does not matter what objects are chosen as primitive signs, and we shall assign natural numbers to this use.[7] Conse-quently, a formula will be a finite sequence of natural numbers,[8] and a proof array a finite sequence of finite sequences of natural numbers. The metamathematical notions (propositions) thus become notions (propositions) about natural numbers or sequences of them;[9] therefore they can (at least in part) be expressed by the symbols of the system *PM* itself. In particular, it can be shown that the notions 'formula', 'proof array', and 'provable formula' can be defined in the system *PM*; that is, we can, for example, find a formula $F(v)$ of *PM* with one free variable $v$ (of the type of a number sequence)[10] such that $F(v)$, interpreted according to the meaning of the terms of *PM*, says: $v$ is a provable formula. We now construct an undecidable proposition of the system *PM*, that is, a proposition $A$ for which neither $A$ nor *not-A* is provable, in the following manner.

A formula of *PM* with exactly one free variable, that variable being of the type of the natural numbers (class of classes), will be called a *class sign*. We assume that the class signs have been arranged in a sequence in some way,[11] we denote the $n$th one by $R(n)$, and we observe that the notion 'class sign', as well as the ordering relation $R$, can be defined in the system *PM*. Let $\alpha$ be any class sign; by $[\alpha; n]$ we denote the formula that results from the class sign $\alpha$ when the free variable is replaced by the sign denoting the natural number $n$. The ternary relation $x = [y; z]$, too, is seen to be definable in *PM*. We now define a class $K$ of natural numbers in the following way:

$$n \, \varepsilon \, K \equiv \overline{Bew} \, [R(n); n] \tag{1}$$

(where *Bew x* means: $x$ is a provable formula.[11a] Since the notions that occur in the definiens can all be defined in *PM*, so can the notion $K$ formed from them; that is, there is a class sign $S$ such that the formula $[S; n]$, interpreted according to the meaning of the terms of *PM*, states that the natural number $n$

belongs to $K$.[12] Since $S$ is a class sign, it is identical with some $R(q)$; that is, we have

$$S = R(q)$$

for a certain natural number $q$. We now show that the proposition $[R(q); q]$ is undecidable in $PM$.[13] For let us suppose that the proposition $[R(q); q]$ were provable; then it would also be true. But in that case, according to the definitions given above, $q$ would belong to $K$, that is, by (1), $Bew[R(q); q]$ would hold, which contradicts the assumption. If, on the other hand, the negation of $[R(q); q]$ were provable, then $\bar{q} \varepsilon K$,[13a] that is, $Bew$ $[R(q); q]$, would hold. But then $[R(q); q]$, as well as its negation, would be provable, which again is impossible.

The analogy of this argument with the Richard antinomy leaps to the eye. It is closely related to the 'Liar' too;[14] for the undecidable proposition $[R(q); q]$ states that $q$ belongs to $K$, that is, by (1), that $[R(q); q]$ is not provable. We therefore have before us a proposition that says about itself that it is not provable [in $PM$].[15] The method of proof just explained can clearly be applied to any formal system that, first, when interpreted as representing a system of notions and propositions, has at its disposal sufficient means of expression to define the notions occurring in the argument above (in particular, the notion 'provable formula') and in which, second, every provable formula is true in the interpretation considered. The purpose of carrying out the above proof with full precision in what follows is, among other things, to replace the second of the assumptions just mentioned by a purely formal and much weaker one.

From the remark that $[R(q); q]$ says about itself that it is not provable it follows at once that $[R(q); q]$ is true, for $[R(q); q]$ *is* indeed unprovable (being undecidable). Thus, the proposition that is undecidable *in the system PM* still was decided by metamathematical considerations. The precise analysis of this curious situation leads to surprising results concerning consistency proofs for formal systems, results that will be discussed in more detail in Section 4 (Theorem XI).

**2.**

We now proceed to carry out with full precision the proof

sketched above. First we give a precise description of the formal system $P$ for which we intend to prove the existence of undecidable propositions. $P$ is essentially the system obtained when the logic of $PM$ is superposed upon the Peano axioms[16] (with the numbers as individuals and the successor relation as primitive notion).

The primitive signs of the system $P$ are the following:

I. Constants: '$\sim$' (not), '$\vee$' (or), '$\Pi$' (for all), '0' (zero), '$f$' (the successor of), '$($,$)$' (parentheses);

II. Variables of type 1 (for individuals, that is, natural numbers including 0): '$x_1$', '$y_1$', '$z_1$', ...;

Variables of type 2 (for classes of individuals): '$x_2$', '$y_2$', '$z_2$', ...;

Variables of type 3 (for classes of classes of individuals): '$x_3$', '$y_3$', '$z_3$', ...;

And so on, for every natural number as a type.[17]

Remark: Variables for functions of two or more argument places (relations) need not be included among the primitive signs since we can define relations to be classes of ordered pairs, and ordered pairs to be classes of classes; for example, the ordered pair $a$, $b$ can be defined to be $((a), (a, b))$, where $(x, y)$ denotes the class whose sole elements are $x$ and $y$, and $(x)$ the class whose sole element is $x$.[18]

By a *sign of type* 1 we understand a combination of signs that has [any one of] the forms

$a$, $fa$, $ffa$, $fffa$, ..., and so on,

where $a$ is either 0 or a variable of type 1. In the first case, we call such a sign a *numeral*. For $n > 1$ we understand by a *sign of type n* the same thing as by a *variable of type n*. A combination of signs that has the form $a(b)$, where $b$ is a sign of type $n$ and $a$ a sign of type $n + 1$, will be called an *elementary formula*. We define the class of *formulas* to be the smallest class[19] containing all elementary formulas and containing $\sim$ $(a)$, $(a) \vee (b)$, $x\Pi(a)$ (where $x$ may be any variable)[19a] whenever it contains $a$ and $b$. We call $(a) \vee (b)$ the *disjunction* of $a$ and $b$, $\sim (a)$ the *negation* and $x\Pi(a)$ a *generalization* of $a$. A formula in which no free variable occurs (*free variable* being defined in the well-known manner) is called a *sentential formula* [Satzformel]. A formula with exactly $n$ free individual variables (and no other free variables) will be called an *n-place relation sign*; for $n = 1$ it will also be called a *class sign*.

By Subst $a\binom{v}{b}$ (where $a$ stands for a formula, $v$ for a variable, and $b$ for a sign of the same type as $v$) we understand the formula that results from $a$ if in $a$ we replace $v$, wherever it is free, by $b$.[20] We say that a formula $a$ is a *type elevation* of another formula $b$ if $a$ results from $b$ when the type of each variable occurring in $b$ is increased by the same number.

The following formulas (I-V) are called *axioms* (we write them using these abbreviations, defined in the well-known manner: ., ⊃, ≡, $(Ex)$, =,[21] and observing the usual conventions about omitting parentheses):[22]

I.   1.  $\sim (fx_1 = 0)$,
     2.  $fx_1 = fy_1 \supset x_1 = y_1$,
     3.  $x_2(0).x_1 \Pi(x_2(x_1) \supset x_2(fx_1)) \supset x_1 \Pi(x_2(x_1))$.

II. All formulas that result from the following schemata by substitution of any formulas whatsoever for $p, q, r$:

1. $p \vee p \supset p$,       3. $p \vee q \supset q \vee p$,
2. $p \supset p \vee q$,       4. $(p \supset q) \supset (r \vee p \supset r \vee q)$.

III. Any formula that results from either one of the two schemata

1. $v\Pi(a) \supset$ Subst $a\binom{v}{c}$,
2. $v\Pi(b \vee a) \supset b \vee v\Pi(a)$

when the following substitutions are made for $a, v, b$, and $c$ (and the operation indicated by 'Subst' is performed in 1):

For $a$ any formula, for $v$ any variable, for $b$ any formula in which $v$ does not occur free, and for $c$ any sign of the same type as $v$, provided $c$ does not contain any variable that is bound in $a$ at a place where $v$ is free.[23]

IV. Every formula that results from the schema

1. $(Eu)(v\Pi(u(v) \equiv a))$

when for $v$ we substitute any variable of type $n$, for $u$ one of type $n + 1$, and for $a$ any formula that does not contain $u$ free. This axiom plays the role of the axiom of reducibility (the comprehension axiom of set theory).

21

V. Every formula that results from

1. $x_1 \varPi(x_2(x_1) \equiv y_2(x_1)) \supset x_2 = y_2$

by type elevation (as well as this formula itself). This axiom states that a class is completely determined by its elements.

A formula $c$ is called an *immediate consequence* of $a$ and $b$ if $a$ is the formula $({\sim}(b)) \vee (c)$, and it is called an *immediate consequence* of $a$ if it is the formula $v\varPi(a)$, where $v$ denotes any variable. The class of *provable formulas* is defined to be the smallest class of formulas that contains the axioms and is closed under the relation 'immediate consequence'.[24]

We now assign natural numbers to the primitive signs of the system $P$ by the following one-to-one correspondence:

| '0' ... 1 | '$\sim$' ... 5 | '$\varPi$' ... 9 |
|-----------|----------------|------------------|
| '$f$' ... 3 | '$\vee$' ... 7 | '(' ... 11 |
| | | ')' ... 13; |

to the variables of type $n$ we assign the numbers of the form $p^n$ (where $p$ is a prime number $> 13$). Thus we have a one-to-one correspondence by which a finite sequence of natural numbers is associated with every finite sequence of primitive signs (hence also with every formula). We now map the finite sequences of natural numbers on natural numbers (again by a one-to-one correspondence), associating the number $2^{n_1}.3^{n_2} \ldots p_k^{n_k}$, where $p_k$ denotes the $k$th prime number (in order of increasing magnitude), with the sequence $n_1, n_2, \ldots, n_k$. A natural number [out of a certain subset] is thus assigned one-to-one not only to every primitive sign but also to every finite sequence of such signs. We denote by $\varPhi(a)$ the number assigned to the primitive sign (or to the sequence of primitive signs) $a$. Now let some relation (or class) $R(a_1, a_2, \ldots, a_n)$ between [or of] primitive signs or sequences of primitive signs be given. With it we associate the relation (or class) $R'(x_1, x_2, \ldots, x_n)$ between [or of] natural numbers that obtains between $x_1, x_2, \ldots, x_n$ if and only if there are some $a_1, a_2, \ldots, a_n$ such that $x_i = \varPhi(a_i)$ ($i = 1, 2, \ldots, n$) and $R(a_1, a_2, \ldots, a_n)$ hold. The relations between (or classes of) natural numbers that in this manner are associated with the metamathematical notions defined so far, for example, 'variable', 'formula', 'sentential formula', 'axiom', 'provable formula', and so on, will be denoted by the same words in SMALL

CAPITALS. The proposition that there are undecidable problems in the system *P*, for example, reads thus: There are SENTENTIAL FORMULAS *a* such that neither *a* nor the NEGATION of *a* is a PROVABLE FORMULA.

We now insert a parenthetic consideration that for the present has nothing to do with the formal system *P*. First we give the following definition: A number-theoretic function[25] $\varphi(x_1, x_2, \ldots, x_n)$ is said to be *recursively defined in terms of* the number-theoretic functions $\psi(x_1, x_2, \ldots, x_{n-1})$ and $\mu(x_1, x_2, \ldots, x_{n+1})$ if

$$\varphi(0, x_2, \ldots, x_n) = \psi(x_2, \ldots, x_n), \qquad (2)$$
$$\varphi(k+1, x_2, \ldots, x_n) = \mu(k, \varphi(k, x_2, \ldots, x_n), x_2, \ldots, x_n)$$

hold for all $x_2, \ldots, x_n, k$.[26]

A number-theoretic function $\varphi$ is said to be *recursive* if there is a finite sequence of number-theoretic functions $\varphi_1, \varphi_2, \ldots, \varphi_n$ that ends with $\varphi$ and has the property that every function $\varphi_k$ of the sequence is recursively defined in terms of two of the preceding functions, or results from any of the preceding functions by substitution,[27] or, finally, is a constant or the successor function $x + 1$. The length of the shortest sequence of $\varphi_i$ corresponding to a recursive function $\varphi$ is called its *degree*. A relation $R(x_1, \ldots, x_n)$ between natural numbers is said to be recursive[28] if there is a recursive function $\varphi(x_1, \ldots, x_n)$ such that, for all $x_1, x_2, \ldots, x_n$,

$$R(x_1, \ldots, x_n) \sim [\varphi(x_1, \ldots, x_n) = 0].[29]$$

The following theorems hold:

I. *Every function (relation) obtained from recursive functions (relations) by substitution of recursive functions for the variables is recursive; so is every function obtained from recursive functions by recursive definition according to schema (2);*

II. *If R and S are recursive relations, so are $\bar{R}$ and $R \vee S$ (hence also R & S);*

III. *If the functions $\varphi(\mathfrak{r})$ and $\psi(\mathfrak{n})$ are recursive, so is the relation $\varphi(\mathfrak{r}) = \psi(\mathfrak{n})$;[30]*

IV. *If the function $\varphi(\mathfrak{r})$ and the relation $R(x, \mathfrak{n})$ are recursive, so are the relations S and T defined by*

$$S(\mathfrak{r}, \mathfrak{n}) \sim (Ex)[x \leq \varphi(\mathfrak{r}) \ \& \ R(x, \mathfrak{n})]$$

*and*

$$T(\mathfrak{r},\mathfrak{n}) \sim (x)[x \leq \varphi(\mathfrak{r}) \rightarrow R(x,\mathfrak{n})],$$

*as well as the function $\psi$ defined by*

$$\psi(\mathfrak{r},\mathfrak{n}) = \varepsilon x[x \leq \varphi(\mathfrak{r}) \,\&\, R(x,\mathfrak{n})],$$

where $\varepsilon x F(x)$ means the least number $x$ for which $F(x)$ holds and 0 in case there is no such number.

Theorem I follows at once from the definition of 'recursive'. Theorems II and III are consequences of the fact that the number-theoretic functions

$$\alpha(x), \quad \beta(x, y), \quad \gamma(x, y),$$

corresponding to the logical notions $\overline{\phantom{--}}$, $\vee$, and $=$, namely,

$\alpha(0) = 1, \alpha(x) = 0 \quad$ for $x \neq 0$,

$\beta(0, x) = \beta(x, 0) = 0, \quad \beta(x, y) = 1 \quad$ when $x$ and $y$ are both $\neq 0$,

$\gamma(x, y) = 0 \quad$ when $x = y$, $\quad \gamma(x, y) = 1 \quad$ when $x \neq y$,

are recursive, as we can readily see. The proof of Theorem IV is briefly as follows. By assumption there is a recursive $\rho(x,\mathfrak{n})$ such that

$$R(x,\mathfrak{n}) \sim [\rho(x,\mathfrak{n}) = 0].$$

We now define a function $\chi(x,\mathfrak{n})$ by the recursion schema (2) in the following way:

$\chi(0,\mathfrak{n}) = 0$,

$\chi(n+1,\mathfrak{n}) = (n+1).a + \chi(n,\mathfrak{n}).\alpha(a),$[31]

where $a = \alpha[\alpha(\rho(0,\mathfrak{n}))].\alpha[\rho(n+1,\mathfrak{n})].\alpha[\chi(n,\mathfrak{n})]$. Therefore $\chi(n+1,\mathfrak{n})$ is equal either to $n+1$ (if $a = 1$) or to $\chi(n,\mathfrak{n})$ (if $a = 0$).[32] The first case clearly occurs if and only if all factors of $a$ are 1, that is, if

$$\bar{R}(0,\mathfrak{n}) \,\&\, R(n+1,\mathfrak{n}) \,\&\, [\chi(n,\mathfrak{n}) = 0]$$

holds. From this it follows that the function $\chi(n, \mathfrak{n})$ (considered as a function of $n$) remains 0 up to [but not including] the least value of $n$ for which $R(n, \mathfrak{n})$ holds and, from there on, is equal to that value. (Hence, in case $R(0, \mathfrak{n})$ holds, $\chi(n, \mathfrak{n})$ is constant and equal to 0.) We have, therefore,

$$\psi(\mathfrak{r}, \mathfrak{n}) = \chi(\varphi(\mathfrak{r}), \mathfrak{n}),$$
$$S(\mathfrak{r}, \mathfrak{n}) \sim R[\psi(\mathfrak{r}, \mathfrak{n}), \mathfrak{n}].$$

The relation $T$ can, by negation, be reduced to a case analogous to that of $S$. Theorem IV is thus proved.

The functions $x + y$, $x \cdot y$, and $x^y$, as well as the relations $x < y$ and $x = y$, are recursive, as we can readily see. Starting from these notions, we now define a number of functions (relations) 1-45, each of which is defined in terms of preceding ones by the procedures given in Theorems I-IV. In most of these definitions several of the steps allowed by Theorems I-IV are condensed into one. Each of the functions (relations) 1-45, among which occur, for example, the notions 'FORMULA', 'AXIOM', and 'IMMEDIATE CONSEQUENCE', is therefore recursive.

1. $x/y \equiv (Ez)[z \leq x \,\&\, x = y.z]$,[33]
$x$ is divisible by $y$.[34]

2. $\mathrm{Prim}(x) \equiv \overline{(Ez)}[z \leq x \,\&\, z \neq 1 \,\&\, z \neq x \,\&\, x/z] \,\&\, x > 1$,
$x$ is a prime number.

3. $0 \, Pr \, x \equiv 0$,
$(n+1) \, Pr \, x \equiv \varepsilon y[y \leq x \,\&\, \mathrm{Prim}(y) \,\&\, x/y \,\&\, y > n \, Pr \, x]$,
$n \, Pr \, x$ is the $n$th prime number (in order of increasing magnitude) contained in $x$.[34a]

4. $0! \equiv 1$,
$(n+1)! \equiv (n+1).n!$.

5. $Pr(0) \equiv 0$,
$Pr(n+1) \equiv \varepsilon y[y \leq \{Pr(n)\}! + 1 \,\&\, \mathrm{Prim}(y) \,\&\, y > Pr(n)]$,
$Pr(n)$ is the $n$th prime number (in order of increasing magnitude).

6. $n \, Gl \, x \equiv \varepsilon y[y \leq x \,\&\, x/(n \, Pr \, x)^y \,\&\, \overline{x/(n \, Pr \, x)^{y+1}}]$,
$n \, Gl \, x$ is the $n$th term of the number sequence assigned to the number $x$ (for $n > 0$ and $n$ not greater than the length of this sequence).

7. $l(x) \equiv \varepsilon y[y \leq x \,\&\, y \, Pr \, x > 0 \,\&\, (y+1) \, Pr \, x = 0]$,
$l(x)$ is the length of the number sequence assigned to $x$.

8. $x*y \equiv \varepsilon z\{z \leq [Pr(l(x) + l(y))]^{x+y} \,\& \,(n)[n \leq l(x) \to$
$\qquad n\,Gl\,z = n\,Gl\,x] \,\& \,(n)[0 < n \leq l(y) \to$
$\qquad (n + l(x))\,Gl\,z = n\,Gl\,y]\}$,

$x*y$ corresponds to the operation of 'concatenating' two finite number sequences.

9. $R(x) \equiv 2^x$,

$R(x)$ corresponds to the number sequence consisting of $x$ alone (for $x > 0$).

10. $E(x) \equiv R(11)*x*R(13)$,

$E(x)$ corresponds to the operation of 'enclosing within parentheses' (11 and 13 are assigned to the primitive signs '(' and ')', respectively).

11. $n\,Var\,x \equiv (Ez)[13 < z \leq x \,\& \,\text{Prim}(z) \,\& \,x = z^n] \,\& \,n \neq 0$,

$x$ is a VARIABLE OF TYPE $n$.

12. $Var(x) \equiv (En)[n \leq x \,\& \,n\,Var\,x]$,

$x$ is a VARIABLE.

13. $\text{Neg}(x) \equiv R(5)*E(x)$,

$\text{Neg}(x)$ is the NEGATION of $x$.

14. $x\,Dis\,y \equiv E(x)*R(7)*E(y)$,

$x\,Dis\,y$ is the DISJUNCTION of $x$ and $y$.

15. $x\,Gen\,y \equiv R(x)*R(q)*E(y)$,

$x\,Gen\,y$ is the GENERALIZATION of $y$ with respect to the VARIABLE $x$ (provided $x$ is a VARIABLE).

16. $0\,N\,x \equiv x$,
$\qquad (n + 1)\,N\,x \equiv R(3)*n\,N\,x$,

$n\,N\,x$ corresponds to the operation of "putting the sign '$f$' $n$ times in front of $x$".

17. $Z(n) \equiv n\,N[R(1)]$,

$Z(n)$ is the NUMERAL denoting the number $n$.

18. $\text{Typ}_1'(x) \equiv (Em, n)\{m, n \leq x \,\& \,[m = 1 \vee 1\,Var\,m] \,\& \,x = n\,N[R(m)]\}$,[34b]

$x$ is a SIGN OF TYPE 1.

19. $\text{Typ}_n(x) \equiv [n = 1 \,\& \,\text{Typ}_1'(x)] \vee [n > 1 \,\& \,(Ev)\{v \leq x \,\& \,n\,Var\,v \,\& \,x = R(v)\}]$,

$x$ is a SIGN OF TYPE $n$.

20. $Elf(x) \equiv (Ey, z, n)[y, z, n \leq x \,\& \,\text{Typ}_n(y) \,\& \,\text{Typ}_{n+1}(z) \,\& \,x = z*E(y)]$,

$x$ is an ELEMENTARY FORMULA.

21. $Op(x, y, z) \equiv x = \text{Neg}(y) \vee x = y\,Dis\,z \vee (Ev)[v \leq x \,\& \,Var(v) \,\& \,x = v\,Gen\,y]$.

22. $FR(x) \equiv (n)\{0 < n \leq l(x) \to Elf(n\,Gl\,x) \vee (Ep,$

$$q)[0 < p, q < n \ \& \ Op(n \ Gl \ x, p \ Gl \ x, q \ Gl \ x)]\}$$
$$\& \ l(x) > 0.$$

$x$ is a SEQUENCE OF FORMULAS, each of which either is an ELE-
MENTARY FORMULA or results from the preceding FORMULAS
through the operations of NEGATION, DISJUNCTION, or GEN-
ERALIZATION.

23. $\text{Form}(x) \equiv (En)\{ \ n \leq (Pr[l(x)^2])^{x \cdot [l(x)]^2} \ \& \ FR(n) \ \& \ x$
$= [l(n)] \ Gl \ n\},$[35]

$x$ is a FORMULA (that is, the last term of a FORMULA SEQUENCE
$n$).

24. $v \ \text{Geb} \ n, x \equiv \text{Var}(v) \ \& \ \text{Form}(x) \ \& \ (Ea, b, c)[a, b, c \leq$
$x \ \& \ x = a * (v \ \text{Gen} \ b) * c \ \& \ \text{Form}(b) \ \& \ l(a) + 1 \leq n \leq$
$l(a) + l(v \ \text{Gen} \ b)],$

the VARIABLE $v$ is BOUND in $x$ at the $n$th place.

25. $v \ \underline{Fr \ n, x} \equiv \text{Var}(v) \ \& \ \text{Form}(x) \ \& \ v = n \ Gl \ x \ \& \ n \leq l(x)$
$\& \ v \ \text{Geb} \ n, x,$

the VARIABLE $v$ is FREE in $x$ at the $n$th place.

26. $v \ Fr \ x \equiv (En)[n \leq l(x) \ \& \ v \ Fr \ n, x],$

$v$ occurs as a FREE VARIABLE in $x$.

27. $Su \ x\binom{n}{y} \equiv \varepsilon z\{ z \leq [Pr(l(x) + l(y))]^{x+y} \ \& \ [(Eu, v) \ u, v$
$\leq x \ \& \ x = u * R(n \ Gl \ x) * v \ \& \ z = u * y * v \ \& \ n$
$= l(u) + 1]\},$

$Su \ x\binom{n}{y}$ results from $x$ when we substitute $y$ for the $n$th term of $x$
(provided that $0 < n \leq l(x)$).

28. $0 \ St \ v, x \equiv \varepsilon n\{n \leq l(x) \ \& \ v \ Fr \ n, x \ \& \ \overline{(Ep)}[n < p \leq$
$l(x) \ \& \ v \ Fr \ \underline{p, x}]\}, (k+1) \ St \ v, x \equiv \varepsilon n\{n < k \ St \ v, x \ \&$
$v \ Fr \ n, x \ \& \ (Ep)[n < p < k \ St \ v, x \ \& \ v \ Fr \ p, x]\},$

$k \ St \ v, x$ is the $(k+1)$th place in $x$ (counted from the right end
of the FORMULA $x$) at which $v$ is FREE in $x$ (and 0 in case there
is no such place).

29. $A(v, x) \equiv \varepsilon n\{n \leq l(x) \ \& \ n \ St \ v, x = 0\},$

$A(v, x)$ is the number of places at which $v$ is FREE in $x$.

30. $Sb_0(x_y^v) \equiv x,$
$Sb_{k+1}(x_y^v) \equiv Su[Sb_k(x_y^v)](\substack{k \ St \ v, \ x \\ y}).$

31. $Sb(x_y^v) \equiv Sb_{A(v,x)}(x_y^v),$[36]

$Sb(x_y^v)$ is the notion SUBST $a\binom{v}{b}$ defined above.[37]

32. $x \ \text{Imp} \ y \equiv [\text{Neg}(x)] \ \text{Dis} \ y,$
$x \ \text{Con} \ y \equiv \text{Neg}\{[\text{Neg}(x)] \ \text{Dis} \ [\text{Neg}(y)]\},$
$x \ \text{Aeq} \ y \equiv (x \ \text{Imp} \ y) \ \text{Con} \ (y \ \text{Imp} \ x),$
$v \ \text{Ex} \ y \equiv \text{Neg}\{v \ \text{Gen} \ [\text{Neg}(y)]\}.$

33. $n \ Th \ x \equiv \varepsilon y\{y \leq x^{(x^n)} \ \& \ (k)[k \leq l(x)$
$\rightarrow (k \ Gl \ x \leq 13 \ \& \ k \ Gl \ y = k \ Gl \ x) \lor (k \ Gl \ x > 13 \ \&$

27

$$k \, Gl \, y = k \, Gl \, x . [1 \, Pr (k \, Gl \, x)]^n)]\},$$

$n \, Th \, x$ is the $n$th TYPE ELEVATION of $x$ (in case $x$ and $n \, Th \, x$ are FORMULAS).

Three specific numbers, which we denote by $z_1$, $z_2$, and $z_3$, correspond to the Axioms I, 1-3, and we define

34. $Z\text{-}Ax(x) \equiv (x = z_1 \lor x = z_2 \lor x = z_3)$.

35. $A_1\text{-}Ax(x) \equiv (Ey)[y \leq x \, \& \, \text{Form}(y) \, \& \, x$
$= (y \, \text{Dis} \, y) \, \text{Imp} \, y]$,

$x$ is a FORMULA resulting from Axiom schema II, 1 by substitution. Analogously, $A_2\text{-}Ax$, $A_3\text{-}Ax$, and $A_4\text{-}Ax$ are defined for Axioms [rather, Axiom Schemata] II, 2-4.

36. $A\text{-}Ax(x) \equiv A_1\text{-}Ax(x) \lor A_2\text{-}Ax(x) \lor A_3\text{-}Ax(x) \lor$
$A_4\text{-}Ax(x)$,

$x$ is a FORMULA resulting from a propositional axiom by substitution.

37. $Q(z, y, v) \equiv \overline{(En, m, w)}[n \leq l(y) \, \& \, m \leq l(z)$
$\& \, w \leq z \, \& \, w = m \, Gl \, z \, \& \, w \, \text{Geb} \, n, y \, \& \, v \, Fr \, n, y.]$

$z$ does not contain any VARIABLE BOUND in $y$ at a place at which $v$ is FREE.

38. $L_1\text{-}Ax(x) \equiv (Ev, y, z, n)\{v, y, z, n \leq x$
$\& \, n \, \text{Var} \, v \, \& \, \text{Typ}_n(z) \, \& \, \text{Form}(y) \, \& \, Q(z, y, v) \, \& \, x$
$= (v \, \text{Gen} \, y) \, \text{Imp} \, [Sb(y_z^v)]\}$,

$x$ is a FORMULA resulting from Axiom schema III, 1 by substitution.

39. $L_2\text{-}Ax(x) \equiv \underline{(Ev, q, p)}\{v, q, p \leq x \, \& \, \text{Var}(v)$
$\& \, \text{Form} \, (p) \, \& \, v \, Fr \, p \, \& \, \text{Form}(q) \, \& \, x$
$= [v \, \text{Gen} \, (p \, \text{Dis} \, q)] \, \text{Imp} \, [p \, \text{Dis} \, (v \, \text{Gen} \, q)]\}$,

$x$ is a FORMULA resulting from Axiom schema III, 2 by substitution.

40. $R\text{-}Ax(x) \equiv (Eu, v, \underline{y, n})[u, v, y, n \leq x \, \& \, n \, \text{Var} \, v$
$\& \, (n + 1) \, \text{Var} \, u \, \& \, u \, Fr \, y \, \& \, \text{Form}(y) \, \& \, x$
$= u \, \text{Ex}\{v \, \text{Gen} \, [[R(u) * E(R(v))] \, \text{Aeq} \, y]\}\}$,

$x$ is a FORMULA resulting from Axiom schema IV, 1 by substitution.

A specific number $z_4$ corresponds to Axiom V, 1, and we define:

41. $M\text{-}Ax(x) \equiv (En)[n \leq x \, \& \, x = n \, Th \, z_4]$.

42. $Ax(x) \equiv Z\text{-}Ax(x) \lor A\text{-}Ax(x) \lor L_1\text{-}Ax(x) \lor$
$L_2\text{-}Ax(x) \lor R\text{-}Ax(x) \lor M\text{-}Ax(x)$,

$x$ is an AXIOM.

43. $Fl(x, y, z) \equiv y = z \, \text{Imp} \, x \lor (Ev)[v \leq x \, \& \, \text{Var}(v)$
$\& \, x = v \, \text{Gen} \, y]$,

$x$ is an IMMEDIATE CONSEQUENCE of $y$ and $z$.

44. $Bw(x) \equiv (n)\{0 < n \leqq l(x) \to Ax(n\,Gl\,x) \lor$
$(Ep, q)[0 < p, q < n \,\&\, Fl(n\,Gl\,x, p\,Gl\,x, q\,Gl\,x)]\}$
$\&\, l(x) > 0,$

$x$ is a PROOF ARRAY (a finite sequence of FORMULAS, each of which is either an AXIOM or an IMMEDIATE CONSEQUENCE of two of the preceding FORMULAS.

45. $x\,B\,y \equiv Bw(x) \,\&\, [l(x)]\,Gl\,x = y,$

$x$ is a PROOF of the FORMULA $y$.

46. $Bew(x) \equiv (Ey)y\,B\,x,$

$x$ is a PROVABLE FORMULA. ($Bew(x)$ is the only one of the notions 1-46 of which we cannot assert that it is recursive.)

The fact that can be formulated vaguely by saying: every recursive relation is definable in the system $P$ (if the usual meaning is given to the formulas of this system), is expressed in precise language, *without* reference to any interpretation of the formulas of $P$, by the following theorem:

Theorem V. *For every recursive relation* $R(x_1, \ldots, x_n)$ *there exists an n-place* RELATION SIGN $r$ (*with the* FREE VARIABLES[38] $u_1, u_2, \ldots, u_n$) *such that for all n-tuples of numbers* $(x_1, \ldots, x_n)$ *we have*

$$R(x_1, \ldots, x_n) \to Bew[Sb(r^{u_1 \ldots u_n}_{Z(x_1) \ldots Z(x_n)})], \tag{3}$$

$$\bar{R}(x_1, \ldots, x_n) \to Bew[Neg(Sb(r^{u_1 \ldots u_n}_{Z(x_1) \ldots Z(x_n)}))]. \tag{4}$$

We shall give only an outline of the proof of this theorem because the proof does not present any difficulty in principle and is rather long.[39] We prove the theorem for all relations $R(x_1, \ldots, x_n)$ of the form $x_1 = \varphi(x_2, \ldots, x_n)$[40] (where $\varphi$ is a recursive function) and we use induction on the degree of $\varphi$. For functions of degree 1 (that is, constants and the function $x + 1$) the theorem is trivial. Assume now that $\varphi$ is of degree $m$. It results from functions of lower degrees, $\varphi_1, \ldots, \varphi_k$, through the operations of substitution or recursive definition. Since by the induction hypothesis everything has already been proved for $\varphi_1, \ldots, \varphi_k$, there are corresponding RELATION SIGNS, $r_1, \ldots, r_k$, such that (3) and (4) hold. The processes of definition by which $\varphi$ results from $\varphi_1, \ldots, \varphi_k$ (substitution and recursive definition) can both be formally reproduced in the system $P$. If this is done, a new RELATION SIGN $r$ is obtained from $r_1, \ldots, r_k$,[41] and, using the induction hypothesis, we can prove without difficulty that

29

(3) and (4) hold for it. A RELATION SIGN $r$ assigned to a recursive relation[42] by this procedure will be said to be recursive.

We now come to the goal of our discussions. Let $\kappa$ be any class of FORMULAS. We denote by $Flg(\kappa)$ (the set of consequences of $\kappa$) the smallest set of FORMULAS that contains all FORMULAS of $\kappa$ and all AXIOMS and is closed under the relation 'IMMEDIATE CONSEQUENCE'. $\kappa$ is said to be $\omega$-consistent if there is no CLASS SIGN $a$ such that

$$(n)[Sb(a_{Z(n)}^{v}) \ \varepsilon \ Flg(\kappa)] \ \& \ [Neg(v \ Gen \ a)] \ \varepsilon \ Flg(\kappa),$$

where $v$ is the FREE VARIABLE of the CLASS SIGN $a$.

Every $\omega$-consistent system, of course, is consistent. As will be shown later, however, the converse does not hold.

The general result about the existence of undecidable propositions reads as follows:

Theorem VI. *For every $\omega$-consistent recursive class $\kappa$ of* FORMULAS *there are recursive* CLASS SIGNS $r$ *such that neither* $v$ Gen $r$ *nor* Neg($v$ Gen $r$) *belongs to* $Flg(\kappa)$ *(where $v$ is the* FREE VARIABLE *of $r$).*

*Proof.* Let $\kappa$ be any recursive $\omega$-consistent class of FORMULAS. We define

$$Bw_\kappa(x) \equiv (n)[n \leq l(x) \rightarrow Ax(n \ Gl \ x) \vee (n \ Gl \ x) \ \varepsilon \ \kappa \ \vee$$
$$(Ep, q)\{0 < p, q < n \ \& \ Fl(n \ Gl \ x, p \ Gl \ x, q \ Gl \ x)\}]$$
$$\& \ l(x) > 0 \tag{5}$$

(see the analogous notion 44),

$$x \ B_\kappa y \equiv Bw_\kappa(x) \ \& \ [l(x)] \ Gl \ x = y \tag{6}$$
$$Bew_\kappa(x) \equiv (Ey)y \ B_\kappa \ x \tag{6.1}$$

(see the analogous notions 45 and 46).
We obviously have

$$(x)[Bew_\kappa(x) \ \sim \ x \ \varepsilon \ Flg(\kappa)] \tag{7}$$

and

$$(x)[Bew(x) \rightarrow Bew_\kappa(x)]. \tag{8}$$

We now define the relation

$$Q(x, y) \equiv \overline{x\,B_{\kappa}\,[Sb(y^{19}_{(y)})]}. \tag{8.1}$$

Since $x\,B_{\kappa}\,y$ (by (6) and (5)) and $Sb(y^{19}_{(y)})$ (by Definitions 17 and 31) are recursive, so is $Q(x, y)$. Therefore, by Theorem V and (8) there is a RELATION SIGN $q$ (with the FREE VARIABLES 17 and 19) such that

$$\overline{x\,B_{\kappa}\,[Sb(y^{19}_{(y)})]} \rightarrow \text{Bew}_{\kappa}[Sb(q^{17}_{Z(x)}\,^{19}_{Z(y)})], \tag{9}$$

and

$$x\,B_{\kappa}[Sb(y^{19}_{(y)})] \rightarrow \text{Bew}_{\kappa}(\text{Neg}(Sb(q^{17}_{Z(x)}\,^{19}_{Z(y)}))]. \tag{10}$$

We put

$$p = 17 \text{ Gen } q \tag{11}$$

($p$ is a CLASS SIGN with the FREE VARIABLE 19) and

$$r = Sb(q^{19}_{Z(p)}) \tag{12}$$

($r$ is a recursive CLASS SIGN[43] with the FREE VARIABLE 17). Then we have

$$Sb(p^{19}_{Z(p)}) = Sb([17 \text{ Gen } q]^{19}_{Z(p)}) = 17 \text{ Gen } Sb(q^{19}_{Z(p)})$$
$$= 17 \text{ Gen } r \tag{13}$$

(by (11) and (12));[44] furthermore

$$Sb(q^{17}_{Z(x)}\,^{19}_{Z(p)}) = Sb(r^{17}_{Z(x)}) \tag{14}$$

(by (12)). If we now substitute $p$ for $y$ in (9) and (10) and take (13) and (14) into account, we obtain

$$\overline{x\,B_{\kappa}\,(17 \text{ Gen } r)} \rightarrow \text{Bew}_{\kappa}[Sb(r^{17}_{Z(x)})], \tag{15}$$

$$x\,B_{\kappa}\,(17 \text{ Gen } r) \rightarrow \text{Bew}_{\kappa}[\text{Neg}(Sb(r^{17}_{Z(x)}))]. \tag{16}$$

This yields:
1. 17 Gen $r$ is not $\kappa$-PROVABLE.[45] For, if it were, there

would (by (6.1)) be an $n$ such that $n \, B_\kappa$ (17 Gen $r$). Hence by (16) we would have $\mathrm{Bew}_\kappa[\mathrm{Neg}(Sb(r^{17}_{Z(n)}))]$, while, on the other hand, from the $\kappa$-PROVABILITY of 17 Gen $r$ that of $Sb(r^{17}_{Z(n)})$ follows. Hence, $\kappa$ would be inconsistent (and a fortiori $\omega$-inconsistent).

2. Neg(17 Gen $r$) is not $\kappa$-PROVABLE. Proof: As has just been proved, 17 Gen $r$ is not $\kappa$-PROVABLE; that is (by (6.1)), $(n)\overline{n \, B_\kappa}$ (17 Gen $r$) holds. From this, $(n)\mathrm{Bew}_\kappa[Sb(r^{17}_{Z(n)})]$ follows by (15), and that, in conjunction with $\mathrm{Bew}_\kappa[\mathrm{Neg}(17$ Gen $r)]$, is incompatible with the $\omega$-consistency of $\kappa$.

17 Gen $r$ is therefore undecidable on the basis of $\kappa$, which proves Theorem VI.

We can readily see that the proof just given is constructive;[45a] that is, the following has been proved in an intuitionistically unobjectionable manner: Let an arbitrary recursively defined class $\kappa$ of FORMULAS be given. Then, if a formal decision (on the basis of $\kappa$) of the SENTENTIAL FORMULA 17 Gen $r$ (which [for each $\kappa$] can actually be exhibited) is presented to us, we can actually give

1. A PROOF of Neg(17 Gen $r$);
2. For any given $n$, a PROOF of $Sb(r^{17}_{Z(n)})$.

That is, a formal decision of 17 Gen $r$ would have the consequence that we could actually exhibit an $\omega$-inconsistency.

We shall say that a relation between (or a class of) natural numbers $R(x_1, \ldots, x_n)$ is *decidable* [*entscheidungsdefinit*] if there exists an $n$-place RELATION SIGN $r$ such that (3) and (4) (see Theorem V) hold. In particular, therefore, by Theorem V every recursive relation is decidable. Similarly, a RELATION SIGN will be said to be *decidable* if it corresponds in this way to a decidable relation. Now it suffices for the existence of undecidable propositions that the class $\kappa$ be $\omega$-consistent and decidable. For the decidability carries over from $\kappa$ to $x \, B_\kappa \, y$ (see (5) and (6) and to $Q(x, y)$ (see (8.1)), and only this was used in the proof given above. In this case the undecidable proposition has the form $v$ Gen $r$, where $r$ is a decidable CLASS SIGN. (Note that it even suffices that $\kappa$ be decidable in the system enlarged by $\kappa$.)

If, instead of assuming that $\kappa$ is $\omega$-consistent, we assume only that it is consistent, then, although the existence of an undecidable proposition does not follow [by the argument given above], it does follow that there exists a property ($r$) for which it is possible neither to give a counterexample nor to prove that it holds of all numbers. For in the proof that 17 Gen $r$ is not $\kappa$-PROVABLE only the consistency of $\kappa$ was used (see pp. 31-2). Moreover from $\text{Bew}_\kappa(17 \text{ Gen } r)$ it follows by (15) that, for every number $x$, $Sb(r^{17}_{Z(x)})$ is $\kappa$-PROVABLE and consequently that $\text{Neg}(SB(r^{17}_{Z(x)}))$ is not $\kappa$-PROVABLE for any number.

If we adjoin Neg(17 Gen $r$) to $\kappa$, we obtain a class of FORMULAS $\kappa'$ that is consistent but not $\omega$-consistent. $\kappa'$ is consistent, since otherwise 17 Gen $r$ would be $\kappa$-PROVABLE. However, $\kappa'$ is not $\omega$-consistent, because, by $\overline{\text{Bew}_\kappa}(17 \text{ Gen } r)$ and (15), $(x(\text{Bew}_\kappa Sb(r^{17}_{Z(x)})$ and, a fortiori, $(x)\text{Bew}_{\kappa'} Sb(r^{17}_{Z(x)}))$ hold, while on the other hand, of course, $\text{Bew}_{\kappa'}[\text{Neg}(17 \text{ Gen } r)]$ holds.[46]

We have a special case of Theorem VI when the class $\kappa$ consists of a finite number of FORMULAS (and, if we so desire, of those resulting from them by TYPE ELEVATION). Every finite class $\kappa$ is, of course, recursive.[46a] Let $a$ be the greatest number contained in $\kappa$. Then we have for $\kappa$

$$x \varepsilon \kappa \sim (Em, n)[m \leq x \& n \leq a \& n \varepsilon \kappa \& x = m \text{ Th } n].$$

Hence $\kappa$ is recursive. This allows us to conclude, for example, that, even with the help of the axiom of choice (for all types) or the generalized continuum hypothesis, not all propositions are decidable, provided these hypotheses are $\omega$-consistent.

In the proof of Theorem VI no properties of the system $P$ were used besides the following:

1. The class of axioms and the rules of inference (that is, the relation 'immediate consequence') are recursively definable (as soon as we replace the primitive signs in some way by natural numbers);

2. Every recursive relation is definable (in the sense of Theorem V) in the system $P$.

Therefore, in every formal system that satisfies the assumptions 1 and 2 and is $\omega$-consistent there are undecidable propositions of the form $(x)F(x)$, where $F$ is a recursively defined property of natural numbers, and likewise in every extension of such a system by a recursively definable $\omega$-consistent class of axioms. As can easily be verified, included among the systems satisfying the assumptions 1 and 2 are the Zermelo-Fraenkel and the von Neumann axiom systems of set

33

theory,[47] as well as the axiom system of number theory consisting of the Peano axioms, recursive definition (by schema (2)), and the rules of logic.[48] Assumption 1 is satisfied by any system that has the usual rules of inference and whose axioms (like those of $P$) result from a finite number of schemata by substitution.[48a]

## 3.

We shall now deduce some consequences from Theorem VI, and to this end we give the following definition:

A relation (class) is said to be *arithmetic* if it can be defined in terms of the notions $+$ and $.$ (addition and multiplication for natural numbers)[49] and the logical constants $\vee$, $\overline{\phantom{--}}$, $(x)$, and $=$, where $(x)$ and $=$ apply to natural numbers only.[50] The notion 'arithmetic proposition' is defined accordingly. The relations 'greater than' and 'congruent modulo $n$', for example, are arithmetic because we have

$$x > y \sim \overline{(Ez)}[y = x + z],$$
$$x \equiv y (\bmod n) \sim (Ez)[x = y + z.n \vee y = x + z.n].$$

We now have

Theorem VII. *Every recursive relation is arithmetic.*

We shall prove the following version of this theorem: every relation of the form $x_0 = \varphi(x_1, \ldots, x_n)$, where $\varphi$ is recursive, is arithmetic, and we shall use induction on the degree of $\varphi$. Let $\varphi$ be of degree $s(s > 1)$. Then we have either

1. $\varphi(x_1, \ldots, x_n) = \rho[\chi_1(x_1, \ldots, x_n), \chi_2(x_1, \ldots, x_n), \ldots, \chi_m(x_1, \ldots, x_n)]^{51}$

(where $\rho$ and all $\chi_1$ are of degrees less than $s$) or

2. $\varphi(0, x_2, \ldots, x_n) = \psi(x_2, \ldots, x_n),$

$$\varphi(k+1, x_2, \ldots, x_n) = \mu[k, \varphi(k, x_2, \ldots, x_n), x_2, \ldots, x_n]$$

(where $\psi$ and $\mu$ are of degrees less than $s$).

In the first case we have

$$x_0 = \varphi(x_1, \ldots, x_n) \sim (Ey_1, \ldots, y_m)[R(x_0, y_1, \ldots, y_m) \;\&\; S_1(y_1, x_1, \ldots, x_n) \;\&\; \ldots \;\&\; S_m(y_m, x_1, \ldots, x_n)],$$

where $R$ and $S_i$ are the arithmetic relations, existing by the induction hypothesis, that are equivalent to $x_0 = \rho(y_1, \ldots, y_m)$ and $y = \chi_i(x_1, \ldots, x_n)$, respectively. Hence in this case $x_0 = \varphi(x_1, \ldots, x_n)$ is arithmetic.

In the second case we use the following method. We can express the relation $x_0 = \varphi(x_1, \ldots, x_n)$ with the help of the notion 'sequence of numbers' $(f)^{52}$ in the following way:

$$x_0 = \varphi(x_1, \ldots, x_n) \sim (Ef)\{f_0 = \psi(x_2, \ldots, x_n)$$
$$\& \ (k)[k < x_1 \to f_{k+1} = \mu(k, f_k, x_2, \ldots, x_n)]$$
$$\& \ x_0 = f_{x_1}\}.$$

If $S(y, x_2, \ldots, x_n)$ and $T(z, x_1, \ldots, x_{n+1})$ are the arithmetic relations, existing by the induction hypothesis, that are equivalent to $y = \psi(x_2, \ldots, x_n)$ and $z = \mu(x_1, \ldots, x_{n+1})$, respectively, then

$$x_0 = \varphi(x_1, \ldots, x_n) \sim (Ef)\{S(f_0, x_2, \ldots, x_n)$$
$$\& \ (k)[k < x_1 \to T(f_{k+1}, k, f_k, x_2, \ldots, x_n)]$$
$$\& \ x_0 = f_{x_1}\}. \tag{17}$$

We now replace the notion 'sequence of numbers' by 'pair of numbers', assigning to the number pair $n$, $d$ the number sequence $f^{(n,d)}$ ($f_k^{(n, \, d)} = [n]_{1+(k+1)d}$), where $[n]_p$ denotes the least nonnegative remainder of $n$ modulo $p$.

We then have

Lemma 1. If $f$ is any sequence of natural numbers and $k$ any natural number, there exists a pair of natural numbers, $n$, $d$ such that $f^{(n, \, d)}$ and $f$ agree in the first $k$ terms.

*Proof.* Let $l$ be the maximum of the numbers $k$, $f_0$, $f_1$, …, $f_{k-1}$. Let us determine an $n$ such that

$$n \equiv f_i[\mathrm{mod}(1 + (i + 1)l!)] \text{ for } i = 0, 1, \ldots, k-1,$$

which is possible, since any two of the numbers $1 + (i + 1)l!$ ($i = 0, 1, \ldots, k - 1$) are relatively prime. For a prime number contained in two of these numbers would also be contained in the difference $(i_1 - i_2)l!$ and therefore, since $|i_1 - i_2| < l$, in $l!$; but this is impossible. The number pair $n$, $l!$ then has the desired property.

Since the relation $x = [n]_p$ is defined by

$$x \equiv n \ (\mathrm{mod} \ p) \ \& \ x < p$$

and is therefore arithmetic, the relation $P(x_0, x_1, \ldots, x_n)$, defined as follows:

$$P(x_0, \ldots, x_n) \equiv (En, d)\{S([n]_{d+1}, x_2, \ldots, x_n)$$
$$\& (k) [k < x_1 \rightarrow T([n]_{1+d(k+2)}, k, [n]_{1+d(k+1)}, x_2, \ldots, x_n)]$$
$$\& x_0 = [n]_{1+d(x_1+1)}\},$$

is also arithmetic. But by (17) and Lemma 1 it is equivalent to $x_0 = \varphi(x_1, \ldots, x_n)$ (the sequence $f$ enters in (17) only through its first $x_1 + 1$ terms). Theorem VII is thus proved.

By Theorem VII, for every problem of the form $(x)F(x)$ (with recursive $F$) there is an equivalent arithmetic problem. Moreover, since the entire proof of Theorem VII (for every particular $F$) can be formalized in the system $P$, this equivalence is provable in $P$. Hence we have

Theorem VIII. *In any of the formal systems mentioned in Theorem VI*[53] *there are undecidable arithmetic propositions.*

By the remark on pages 33-4, the same holds for the axiom system of set theory and its extensions by $\omega$-consistent recursive classes of axioms.

Finally, we derive the following result:

Theorem IX. *In any of the formal systems mentioned in Theorem VI*[53] *there are undecidable problems of the restricted functional calculus*[54] (that is, formulas of the restricted functional calculus for which neither validity nor the existence of a counterexample is provable).[55]

This is a consequence of

Theorem X. *Every problem of the form* $(x)F(x)$ *(with recursive F) can be reduced to the question whether a certain formula of the restricted functional calculus is satisfiable* (that is, for every recursive $F$ we can find a formula of the restricted functional calculus that is satisfiable if and only if $(x)F(x)$ is true.

By formulas of the restricted functional calculus (r.f.c.) we understand expressions formed from the primitive signs $\overline{\phantom{-}}$, $\vee$, $(x)$, $=$, $x$, $y$, $\ldots$ (individual variable), $F(x)$, $G(x, y)$, $H(x, y, z)$, $\ldots$ (predicate and relation variable), where $(x)$ and $=$ apply to individuals only.[56] To these signs we add a third kind of variables, $\varphi(x)$, $\psi(x, y)$, $\kappa(x, y, z)$, and so on, which stand for object-functions [*Gegenstandsfunktionen*] (that is, $\varphi(x)$, $\psi(x, y)$, and so on denote single-valued functions whose arguments and values are individuals).[57] A formula that contains variables of the third kind in addition to the signs of the r.f.c. first mentioned will be called a formula in the extended sense (i.e.s.).[58] The notions 'satisfiable' and 'valid' carry over immediately to formulas i.e.s., and we have the theorem that, for any formula $A$ i.e.s., we can find a formula $B$ of the r.f.c. proper such that $A$ is satisfiable if and only if $B$ is. We obtain $B$ from $A$ by replacing the variables of the third kind, $\varphi(x)$, $\psi(x, y)$, ..., that occur in $A$ with expressions of the form $(\iota z)F(z, x)$, $(\iota z)G(z, x, y)$, ..., by eliminating the 'descriptive' functions by the method used in $PM$ (I, *14), and by logically multiplying[59] the formula thus obtained by an expression stating about each $F$, $G$, ... put in place of some $\varphi$, $\psi$, ... that it holds for a unique value of the first argument [for any choice of values for the other arguments].

We now show that, for every problem of the form $(x)F(x)$ (with recursive $F$), there is an equivalent problem concerning the satisfiability of a formula i.e.s., so that, on account of the remark just made, Theorem X follows.

Since $F$ is recursive, there is a recursive function $\Phi(x)$ such that $F(x) \sim [\Phi(x) = 0]$, and for $\Phi$ there is a sequence of functions, $\Phi_1$, $\Phi_2$, ... $\Phi_n$, such that $\Phi_n$, $= \Phi$, $\Phi_1(x) = x + 1$, and for every $\Phi_k$ $(1 < k \le n)$ we have either

1.  $(x_2, ..., x_m)[\Phi_k(0, x_2, ..., x_m) = \Phi_p(x_2, ..., x_m)]$,
    $(x, x_2, ..., x_m)\{\Phi_k(\Phi_1(x), x_2, ..., x_m)$
    $= \Phi_q[x, \Phi_k(x, x_2, ..., x_m), x_2, ..., x_m]\}$, (18)
    with $p, q < k$,[59a]

or

2.  $(x_1, ..., x_m)[\Phi_k(x_1, ..., x_m) = \Phi_r(\Phi_{i_1}(\mathfrak{r}_1), ...,$
    $\Phi_{i_s}(\mathfrak{r}_s))]$,[60] (19)
    with $r < k$, $i_\nu < k$ (for $\nu = 1, 2, ..., s$),

or

3.  $(x_1, ..., x_m)[\Phi_k(x_1, ..., x_m)$
    $= \Phi_1(\Phi_1(...(\Phi_1(0))...))]$. (20)

We then form the propositions

$$(x)\overline{\Phi_1(x) = 0} \ \& \ (x, y)[\Phi_1(x) = \Phi_1(y) \to x = y], \qquad (21)$$

$$(x)[\Phi_n(x) = 0]. \tag{22}$$

In all of the formulas (18), (19), (20) (for $k = 2, 3, \ldots, n$) and in (21) and (22) we now replace the functions $\Phi_i$ by function variables $\varphi_i$ and the number 0 by an individual variable $x_0$ not used so far, and we form the conjunction $C$ of all the formulas thus obtained.

The formula $(Ex_0)C$ then has the required property, that is,

1. If $(x)[\Phi(x) = 0]$ holds, $(Ex_0)C$ is satisfiable. For the functions $\Phi_1, \Phi_2, \ldots, \Phi_n$ obviously yield a true proposition when substituted for $\varphi_1, \varphi_2, \ldots, \varphi_n$ in $(Ex_0)C$;

2. If $(Ex_0)C$ is satisfiable, $(x)[\Phi(x) = 0]$ holds.

*Proof.* Let $\psi_1, \psi_2, \ldots, \psi_n$ be the functions (which exist by assumption) that yield a true proposition when substituted for $\varphi_1, \varphi_2, \ldots, \varphi_n$ in $(Ex_0)C$. Let $\mathfrak{J}$ be their domain of individuals. Since $(Ex_0)C$ holds for the functions $\psi_i$, there is an individual $a$ (in $\mathfrak{J}$) such that all of the formulas (18)-(22) go over into true propositions, (18')-(22'), when the $\Phi_i$ are replaced by the $\psi_i$ and 0 by $a$. We now form the smallest subclass of $\mathfrak{J}$ that contains $a$ and is closed under the operation $\psi_1(x)$. This subclass ( $\mathfrak{J}'$ ) has the property that every function $\psi_i$, when applied to elements of $\mathfrak{J}'$, again yields elements of $\mathfrak{J}'$. For this holds of $\psi_1$ by the definition of $\mathfrak{J}'$, and by (18'), (19'), and (20') it carries over from $\psi_i$ with smaller subscripts to $\psi_i$ with larger ones. The functions that result from the $\psi_i$ when these are restricted to the domain $\mathfrak{J}'$ of individuals will be denoted by $\psi_i'$. All of the formulas (18)-(22) hold for these functions also (when we replace 0 by $a$ and $\Phi_i$ by $\psi_i'$).

Because (21) holds for $\psi_1'$ and $a$, we can map the individuals of $\mathfrak{J}'$ one-to-one onto the natural numbers in such a manner that $a$ goes over into 0 and the function $\psi_1'$ into the successor function $\Phi_1$. But by this mapping the functions $\psi_i'$ go over into the functions $\Phi_i$, and, since (22) holds for $\psi_n'$ and $a$, $(x)[\Phi_n(x) = 0]$, that is, $(x)[\Phi(x) = 0]$, holds, which was to be proved.[61]

Since (for each particular $F$) the argument leading to Theorem X can be carried out in the system $P$, it follows that any proposition of the form $(x)F(x)$ (with recursive $F$) can in $P$ be proved equivalent to the proposition that states about the corresponding formula of the r.f.c. that it is satisfiable. Hence the undecidability of one implies that of the other, which proves Theorem IX.[62]

**4.**

The results of Section 2 have a surprising consequence concerning a consistency proof for the system $P$ (and its extensions), which can be stated as follows:

Theorem XI. *Let $\kappa$ be any recursive consistent*[63] *class of* FORMULAS; *then the* SENTENTIAL FORMULA *stating that $\kappa$ is consistent is not $\kappa$-PROVABLE;* in particular, the consistency of $P$ is not provable in $P$,[64] provided $P$ is consistent (in the opposite case, of course, every proposition is provable [in $P$]).

The proof (briefly outlined) is as follows. Let $\kappa$ be some recursive class of FORMULAS chosen once and for all for the following discussion (in the simplest case it is the empty class). As appears from 1, pages 31-2, only the consistency of $\kappa$ was used in proving that 17 Gen $r$ is not $\kappa$-PROVABLE;[65] that is, we have

$$\text{Wid}(\kappa) \rightarrow \overline{\text{Bew}_\kappa}(17 \text{ Gen } r), \tag{23}$$

that is, by (6.1),

$$\text{Wid}(\kappa) \rightarrow (x) \overline{x \, B_\kappa \, (17 \text{ Gen } r)}.$$

By (13), we have

$$17 \text{ Gen } r = Sb(p\,{}^{19}_{Z(p)}),$$

hence

$$\text{Wid}(\kappa) \rightarrow (x) \overline{x \, B_\kappa \, Sb(p\,{}^{19}_{Z(p)})},$$

that is, by (8.1),

$$\text{Wid}(\kappa) \rightarrow (x) Q(x, p). \tag{24}$$

We now observe the following: all notions defined (or statements proved) in Section 2,[66] and in Section 4 up to this point, are also expressible (or provable) in $P$. For throughout we have used only the methods of definition and proof that are customary in classical mathematics, as they are formalized in the system $P$. In particular, $\kappa$ (like every recursive class) is definable in $P$. Let $w$ be the SENTENTIAL FORMULA by which Wid($\kappa$) is expressed in $P$. According to (8.1), (9), and (10), the relation $Q(x, y)$ is expressed by the RELATION SIGN $q$, hence $Q(x, p)$ by $r$ (since, by (12), $r = Sb(q^{19}_{Z(p)})$), and the proposition $(x)Q(x, p)$ by 17 Gen $r$.

Therefore, by (24), $w$ Imp (17 Gen $r$) is provable in $P$[67] (and a fortiori $\kappa$-PROVABLE). If now $w$ were $\kappa$-PROVABLE, then 17 Gen $r$ would also be $\kappa$-PROVABLE, and from this it would follow, by (23), that $\kappa$ is not consistent.

Let us observe that this proof, too, is constructive; that is, it allows us to actually derive a contradiction from $\kappa$, once a PROOF of $w$ from $\kappa$ is given. The entire proof of Theorem XI carries over word for word to the axiom system of set theory, $M$, and to that of classical mathematics,[68] $A$, and here, too, it yields the result: There is no consistency proof for $M$, or for $A$, that could be formalized in $M$, or $A$, respectively, provided $M$, or $A$, is consistent. I wish to note expressly that Theorem XI (and the corresponding results for $M$ and $A$) do not contradict Hilbert's formalistic viewpoint. For this viewpoint presupposes only the existence of a consistency proof in which nothing but finitary means of proof is used, and it is conceivable that there exist finitary proofs that *cannot* be expressed in the formalism of $P$ (or of $M$ or $A$).

Since, for any consistent class $\kappa$, $w$ is not $\kappa$-PROVABLE, there always are propositions (namely $w$) that are undecidable (on the basis of $\kappa$) as soon as Neg($w$) is not $\kappa$-PROVABLE; in other words, we can, in Theorem VI, replace the assumption of $\omega$-consistency by the following: The proposition '$\kappa$ is inconsistent' is not $\kappa$-PROVABLE. (Note that there are consistent $\kappa$ for which this proposition is $\kappa$-PROVABLE.)

In the present paper we have on the whole restricted ourselves to the system $P$, and we have only indicated the applications to other systems. The results will be stated and proved in full generality in a sequel to be published soon.[68a] In that paper, also, the proof of Theorem XI, only sketched here, will be given in detail.

*Note added 28 August 1963.* In consequence of later advances, in particular of the fact that due to A.M. Turing's work[69] a precise and unquestionably adequate definition of the general notion of formal system[70] can now be given, a completely general version of Theorems VI and XI is now possible. That is, it can be proved rigorously that in *every* consistent formal system that contains a certain amount of finitary number theory there exist undecidable arithmetic propositions and that, moreover, the consistency of any such system cannot be proved in the system.

## NOTES

1. See a summary of the results of the present paper in Gödel (1930b).

2. Whitehead and Russell (1925). Among the axioms of the system *PM* we include also the axiom of infinity (in this version: there are exactly denumerably many individuals), the axiom of reducibility, and the axiom of choice (for all types).

3. See Fraenkel (1927) and von Neumann (1925, 1928, 1929). We note that in order to complete the formalization we must add the axioms and rules of inference of the calculus of logic to the set-theoretic axioms given in the literature cited. The considerations that follow apply also to the formal systems (so far as they are available at present) constructed in recent years by Hilbert and his collaborators. See Hilbert (1922, 1922a, 1927), Bernays (1923), von Neumann (1927), and Ackermann (1924).

4. That is, more precisely, there are undecidable propositions in which, besides the logical constants $\overline{\phantom{xx}}$ (not), $\vee$ (or), $(x)$ (for all), and $=$ (identical with), no other notions occur but $+$ (addition) and $.$ (multiplication), both for natural numbers, and in which the prefixes $(x)$, too, apply to natural numbers only.

5. In *PM* only axioms that do not result from one another by mere change of type are counted as distinct.

6. Here and in what follows we always understand by 'formula of *PM*' a formula written without abbreviations (that is, without the use of definitions). It is well known that [in *PM*] definitions serve only to abbreviate notations and therefore are dispensable in principle.

7. That is, we map the primitive signs one-to-one into some natural numbers. (See how this is done on page 22.)

8. That is, a number-theoretic function defined on an initial segment of the natural numbers. (Numbers, of course, cannot be arranged in a spatial order.)

9. In other words, the procedure described above yields an isomorphic image of the system *PM* in the domain of arithmetic, and all metamathematical arguments can just as well be carried out in this isomorphic image. This is what we do below when we sketch the proof; that is, by 'formula', 'proposition', 'variable', and so on, *we must always understand the corresponding objects of the isomorphic image.*

10. It would be very easy (although somewhat cumbersome) to actually write down this formula.

11. For example, by increasing the sum of the finite sequence of integers that is the 'class sign', and lexicographically for equal sums.

11a. The bar denotes negation.

12. Again, there is not the slightest difficulty in actually writing down the formula $S$.

13. Note that '$[R(q); q]$' (or, which means the same, '$[S; q]$') is merely a *metamathematical description* of the undecidable proposition. But, as soon as the formula $S$ has been obtained, we can, of course, also determine the number $q$ and, therewith, actually write down the undecidable proposition itself. [This makes no difficulty in principle. However, in order not to run into formulas of entirely unmanageable lengths and to avoid practical difficulties in the computation of the number $q$, the construction of the undecidable proposition would have to be slightly modified, unless the technique of abbreviation by definition used throughout in $PM$ is adopted.]

13a. [The German text reads $n \, \varepsilon \, K$, which is a misprint.]

14. Any epistemological antinomy could be used for a similar proof of the existence of undecidable propositions.

15. Contrary to appearances, such a proposition involves no faulty circularity, for initially it [only] asserts that a certain well-defined formula (namely, the one obtained from the $q$th formula in the lexicographic order by a certain substitution) is unprovable. Only subsequently (and so to speak by chance) does it turn out that this formula is precisely the one by which the proposition itself was expressed.

16. The addition of the Peano axioms, as well as all other modifications introduced in the system $PM$, merely serves to simplify the proof and is dispensable in principle.

17. It is assumed that we have denumerably many signs at our disposal for each type of variable.

18. Nonhomogeneous relations, too, can be defined in this manner; for example, a relation between individuals and classes can be defined to be a class of elements of the form $((x_2), ((x_1), x_2))$. Every proposition about relations that is provable in $PM$ is provable also when treated in this manner, as is readily seen.

19. Concerning this definition (and similar definitions occurring below) see Lukasiewicz and Tarski (1930).

19a. Hence $x\Pi(a)$ is a formula even if $x$ does not occur in $a$ or is not free in $a$. In this case, of course, $x\Pi(a)$ means the same thing as $a$.

20. In case $v$ does not occur in $a$ as a free variable we put Subst $a\binom{v}{b}$ = $a$. Note that 'Subst' is a metamathematical sign.

21. $x_1 = y_1$ is to be regarded as defined by $x_2\Pi(x_2(x_1) \supset x_2(y_1))$, as in $PM$ (I, *13) similarly for higher types).

22. In order to obtain the axioms from the schemata listed we must therefore

(1) Eliminate the abbreviations and

(2) Add the omitted parentheses

(in II, III, and IV after carrying out the substitutions allowed).

Note that all expressions thus obtained are 'formulas' in the sense specified above. (See also the exact definitions of the metamathematical notions on pp. 24-29.)

23. Therefore $c$ is a variable or 0 or a sign of the form $f \dots fu$, where $u$ is either 0 or a variable of type 1. Concerning the notion 'free (bound) at a place in $a$', see I A 5 in von Neumann (1927).

24. The rule of substitution is rendered superfluous by the fact that all possible substitutions have already been carried out in the axioms themselves. (This procedure was used also in von Neumann (1927).)

25. That is, its domain of definition is the class of nonnegative integers (or of $n$-tuples of nonnegative integers) and its values are nonnegative integers.

26. In what follows, lower-case italic letters (with or without subscripts) are always variables for nonnegative integers (unless the contrary is expressly noted).

27. More precisely, by substitution of some of the preceding functions at the argument places of one of the preceding functions, for example, $\varphi_k(x_1, x_2) = \varphi_p[\varphi_q(x_1, x_2), \varphi_r(x_2)]$ $(p, q, r < k)$. Not all variables on the left side need occur on the right side (the same applies to the recursion schema (2)).

28. We include classes among relations (as one-place relations). Recursive relations $R$, of course, have the property that for every given $n$-tuple of numbers it can be decided whether $R(x_1, \ldots, x_n)$ holds or not.

29. Whenever formulas are used to express a meaning (in particular, in all formulas expressing metamathematical propositions or notions), Hilbert's symbolism is employed. See Hilbert and Ackermann (1928).

30. We use German letters, $\mathfrak{r}$ , $\mathfrak{n}$ , as abbreviations for arbitrary $n$-tuples of variables, for example, $x_1, x_2, \ldots, x_n$.

31. We assume familiarity with the fact that the functions $x + y$ (addition) and $x \cdot y$ (multiplication) are recursive.

32. $a$ cannot take values other than 0 and 1, as can be seen from the definition of $a$.

33. The sign $\equiv$ is used in the sense of 'equality by definition'; hence in definitions it stands for either $=$ or $\sim$ (otherwise, the symbolism is Hilbert's).

34. Wherever one of the signs $(x)$, $(Ex)$, or $\varepsilon x$ occurs in the definitions below, it is followed by a bound on $x$. This bound merely serves to ensure that the notion defined is recursive (see Theorem IV). But in most cases the *extension* of the notion defined would not change if this bound were omitted.

34a. For $0 < n \leq z$, where $z$ is the number of distinct prime factors of $x$. Note that $n\,Pr\,x = 0$ for $n = z + 1$.

34b. $m, n \leq x$ stands for $m \leq x$ & $n \leq x$ (similarly for more than two variables).

35. That $n \leq (Pr([l(x)]^2))^{x \cdot [l(x)]^2}$ provides a bound can be seen thus: The length of the shortest sequence of formulas that corresponds to $x$ can at most be equal to the number of subformulas of $x$. But there are at most $l(x)$ subformulas of length 1, at most $l(x) - 1$ of length 2, and so on, hence altogether at most $l(x)(l(x) + 1)/2 \leq [l(x)]^2$. Therefore all prime factors of $n$ can be assumed to be less than $Pr([l(x)]^2)$, their number $\leq [(lx)]^2$, and their exponents (which are subformulas of $x$) $\leq x$.

36. In case $v$ is not a VARIABLE or $x$ is not a FORMULA, $Sb(x_v^y) = x$.

37. Instead of $Sb[Sb(x_v^y)_w^z]$ we write $Sb(x_{v\,w}^{y\,z})$ (and similarly for more than two VARIABLES).

38. The VARIABLES $u_1, \ldots, u_n$ can be chosen arbitrarily. For example, there always is an $r$ with the FREE VARIABLES 17, 19, 23, ..., and so on, for which (3) and (4) hold.

39. Theorem V, of course, is a consequence of the fact that in the case of a recursive relation $R$ it can, for every $n$-tuple of numbers, be decided *on the basis of the axioms of the system P* whether the relation $R$ obtains or not.

40. From this it follows at once that the theorem holds for every recursive relation, since any such relation is equivalent to $0 = \varphi(x_1, \ldots, x_n)$, where $\varphi$ is recursive.

41. When this proof is carried out in detail, $r$, of course, is not defined indirectly with the help of its meaning but in terms of its purely formal structure.

42. Which, therefore, in the usual interpretation expresses the fact that this relation holds.

43. Since $r$ is obtained from the recursive RELATION SIGN $q$ through the replacement of a VARIABLE by a definite number, $p$. [Precisely stated the final part of this footnote (which refers to a side remark unnecessary for the proof) would read thus: 'REPLACEMENT of a VARIABLE by the NUMERAL for $p$.']

44. The operations Gen and *Sb*, of course, can always be interchanged in case they refer to different VARIABLES.

45. By '$x$ is $\kappa$-*provable*' we mean $x \, \varepsilon \, \mathrm{Flg}(\kappa)$, which, by (7), means the same thing as $\mathrm{Bew}_\kappa(x)$.

45a. Since all existential statements occurring in the proof are based upon Theorem V, which, as is easily seen, is unobjectionable from the intuitionistic point of view.

46. Of course, the existence of classes $\kappa$ that are consistent but not $\omega$-consistent is thus proved only on the assumption that there exists some consistent $\kappa$ (that is, that $P$ is consistent).

46a. [On page 190, lines 21, 22, and 23, of the German text the three occurrences of $\alpha$ are misprints and should be replaced by occurrences of $\kappa$.]

47. The proof of assumption 1 turns out to be even simpler here than for the system $P$, since there is just one kind of primitive variable (or two in von Neumann's system).

48. See Problem III in Hilbert (1928a).

48a. As will be shown in Part II of this paper, the true reason for the incompleteness inherent in all formal systems of mathematics is that the formation of ever higher types can be continued into the transfinite (see Hilbert (1925), p. 184), while in any formal system at most denumerably many of them are available. For it can be shown that the undecidable propositions constructed here become decidable whenever appropriate higher types are added (for example, the type $\omega$ to the system $P$). An analogous situation prevails for the axiom system of set theory.

49. Here and in what follows, zero is always included among the natural numbers.

50. The definiens of such a notion, therefore, must consist exclusively of the signs listed, variables for natural numbers, $x, y, \ldots$, and the

signs 0 and 1 (variables for functions and sets are not permitted to occur). Instead of $x$ any other number variable, of course, may occur in the prefixes.

51. Of course, not all $x_1, \ldots, x_n$ need occur in the $\chi_1$ (see the example in footnote 27).

52. $f$ here is a variable with the [infinite] sequences of natural numbers as its domain of values. $f_k$ denotes the $(k + 1)$th term of a sequence $f$ ($f_0$ denoting the first).

53. These are the $\omega$-consistent systems that result from $P$ when recursively definable classes of axioms are added.

54. See Hilbert and Ackermann (1928).

In the system $P$ we must understand by formulas of the restricted functional calculus those that result from the formulas of the restricted functional calculus of $PM$ when relations are replaced by classes of higher types as indicated on page 20.

55. In 1930 I showed that every formula of the restricted functional calculus either can be proved to be valid or has a counterexample. However, by Theorem IX the existence of this counterexample is *not* always provable (in the formal systems we have been considering).

56. Hilbert and Ackermann (1928) do not include the sign = in the restricted functional calculus. But for every formula in which the sign = occurs there exists a formula that does not contain this sign and is satisfiable if and only if the original formula is (see Gödel, 1930a).

57. Moreover, the domain of definition is always supposed to be the *entire* domain of individuals.

58. Variables of the third kind may occur at all argument places occupied by individual variables, for example, $y = \varphi(x)$, $F(x, \varphi(y))$, $G(\psi(x, \varphi(y)), x)$, and the like.

59. That is, by forming the conjunction.

59a. [The last clause of footnote 27 was not taken into account in the formulas (18). But an explicit formulation of the cases with fewer variables on the right side is actually necessary here for the formal correctness of the proof, unless the identity function, $I(x) = x$, is added to the initial functions.]

60. The $\mathfrak{r}_i$ ($i = 1, \ldots, s$) stand for finite sequences of the variables $x_1, x_2, \ldots, x_m$; for example, $x_1, x_3, x_2$.

61. Theorem X implies, for example, that Fermat's problem and Goldbach's problem could be solved if the decision problem for the r.f.c. were solved.

62. Theorem IX, of course, also holds for the axiom system of set theory and for its extensions by recursively definable $\omega$-consistent classes of axioms, since there are undecidable propositions of the form $(x) F(x)$ (with recursive $F$) in these systems too.

63. '$\kappa$ is consistent' (abbreviated by 'Wid($\kappa$)') is defined thus: Wid($\kappa$) $\equiv$ $(Ex)(\text{Form}(x) \,\&\, \overline{\text{Bew}}_\kappa(x))$.

64. This follows if we substitute the empty class of FORMULAS for $\kappa$.

65. Of course, $r$ (like $p$) depends on $\kappa$.

66. From the definition of 'recursive' on page 23 to the proof of Theorem VI inclusive.

67. That the truth of $w$ Imp ($17$ Gen $r$) can be inferred from (23) is

simply due to the fact that the undecidable proposition 17 Gen $r$ asserts its own unprovability, as was noted at the very beginning.

68. See von Neumann (1927).

68a. [This explains the 'I' in the title of the paper. The author's intention was to publish this sequel in the next volume of the *Monatshefte*. The prompt acceptance of his results was one of the reasons that made him change his plan.]

69. See Turing (1937), p. 249.

70. In my opinion the term 'formal system' or 'formalism' should never be used for anything but this notion. In a lecture at Princeton (mentioned in *Princeton University 1946*, p. 11 [see Davis, 1965, pp. 84-88]) I suggested certain transfinite generalizations of formalisms, but these are something radically different from formal systems in the proper sense of the term, whose characteristic property is that reasoning in them, in principle, can be completely replaced by mechanical devices.

## BIBLIOGRAPHY

Ackermann, Wilhelm (1924) 'Begründung des "tertium non datur" mittels der Hilbertschen Theorie der Wiserspruchsfreiheit', *Mathematische Annalen* 93, 1-36

Bernays, Paul (1923) 'Erwiderung auf die Note von Herrn Aloys Müller: "über Zahlen als Zeichen"', *Mathematische Annalen* 60, 159-63; reprinted in *Annalen der Philosophie und philosophischen Kritik* 4 (1924), 492-7

Davis, Martin (ed.) (1965) *The Undecidable. Basic Papers on Undecidable Propositions, Unsolvable Problems and Computable Functions* (Hewlett, New York, Raven Press)

Fraenkel, Abraham A. (1927) *Zehn Vorlesungen über die Grundlegung der Mengenlehre* (Leipzig and Berlin, Teubner)

Gödel, Kurt (1930b) 'Einige metamathematische Resultate über Entscheidungsdefinitheit und Widerspruchsfreiheit', *Anzeiger der Akademie der Wissenschaften in Wien, Mathematisch-naturwissenschaftliche Klasse* 67, 214-15; reprinted in van Heijenoort (1967)

—— (1930a) 'Über die vollständigkeit des Logikkalküls', *Die Naturwissenschaften*, 18, 1068

Hilbert, David (1922) 'Neubegründung der Mathematik (Erste Mitteilung)', *Abhandlungen aus dem mathematischen Seminar der Hamburgischen Universität* 1, 157-77; reprinted in Hilbert (1935), 157-77

—— (1922a) 'Die Logischen Grundlagen der Mathematik', *Mathematische Annalen* 88 (1923), 151-65; reprinted in Hilbert (1935), 178-91

—— (1925) 'Über das Unendliche', *Mathematische Annalen* 95 (1926); English translation in van Heijenoort (1967, 367-92)

—— (1927) 'Die Grundlagen der Mathematik', *Abhandlungen aus dem mathematischen Seminar der Hamburgischen Universität* 6

(1928), 65-85; reprinted in Hilbert (1928, 1-21 and 1930a, 289-312)

—— (1928) *Die Grundlagen der Mathematik*, mit Zusätzen von Hermann Weyl und Paul Bernays, *Hamburger Mathematische Einzelschriften* 5 (Leipzig, Teubner)

—— (1928a) 'Probleme der Grundlegung der Mathematik', *Atti del Congresso internazionale dei matematici, Bologna 3-10 Settembre 1928* (Bologna, 1929), vol. 1, 135-41; reprinted, with emendations and additions, in *Mathematische Annalen* 102 (1929), 1-9, and Hilbert (1930a, 313-23)

—— (1930a) *Grundlagen der Geometrie*, 7th edn (Leipzig and Berlin, Teubner)

—— (1935) *Gesammelte Abhandlungen* (Berlin, Springer), 3 vols.

—— and Wilhelm Ackermann (1928) *Grundzüge der theoretischen Logik* (Berlin, Springer)

Łukasiewicz, Jan and Alfred Tarski (1930) 'Untersuchungen über den Aussagenkalkül', *Sprawozdania z posiedzeń Towarzystwa Naukowego Warszawskiego, Wydział* III, 23, 30-50; English translation in Tarski (1956, 38-59)

Tarski, Alfred (1956) *Logic, Semantics, Metamathematics* (Oxford, Oxford University Press)

Turing, Alan (1937) 'On computable numbers, with an application to the Entscheidungsproblem', *Proceedings of the London Mathematical Society*, 2nd series, 42, 230-65; correction, *ibid.*, 43, 544-6; reprinted in Davis (1965, 116-54)

van Heijenoort, Jean (ed.) (1967) *From Frege to Gödel: A Source Book in Mathematical Logic, 1879-1931* (Cambridge, Mass., Harvard University Press)

von Neumann, John (1925) 'Eine Axiomatisierung der Mengenlehre', *Journal für die reine und angewandte Mathematik* 154, 219-40, reprinted in von Neumann (1961, 34-56)

—— (1927) 'Zur Hilbertschen Beweistheorie', *Mathematische Zeitschrift* 26, 1-46; reprinted in von Neumann (1961, 256-300)

—— (1928) 'Über die Definition durch transfinite Induktion und verwandte Fragen der allgemeinen Mengenlehre', *Mathematische Annalen* 99, 373-91; reprinted in von Neumann (1961, 320-38)

—— (1929) 'Über eine Widerspruchsfreiheitsfrage in der axiomatischen Mengenlehre', *Journal für die reine und angewandte Mathematik* 160, 227-41; reprinted in von Neumann (1961, 494-508)

—— (1961) *Collected Works*, vol. 1 (New York, Pergamon Press)

Whitehead, Alfred North and Bertrand Russell (1925) *Principia mathematica*, vol. 1, 2nd edn (Cambridge, Cambridge University Press)

# III

# The Work of Kurt Gödel

Stephen C. Kleene[1]

I first heard the name of Kurt Gödel when, as a graduate student at Princeton in the fall of 1931, I attended a colloquium at which John von Neumann was the speaker. Von Neumann could have spoken on work of his own; but instead he gave an exposition of Gödel's results on formally undecidable propositions (1931).

Today I shall begin with Gödel's paper (1930) on *The completeness of the axioms of the functional calculus of logic*, or of what we now often call 'the first-order predicate calculus', using 'predicate' as synonymous with 'propositional function'.

Alonzo Church wrote (1944, p. 62 and 1956, pp. 288-9), 'the first explicit formulation of the functional calculus of first order as an independent logistic system is perhaps in the first edition of Hilbert and Ackermann's *Grundzüge der theoretischen Logik* (1928)'. Clearly, this formalism is not complete in the sense that each closed formula or its negation is provable. (A *closed formula*, or *sentence*, is a formula without free occurrences of variables.) But Hilbert and Ackermann observe, 'Whether the system of axioms is complete at least in the sense that all the logical formulas which are correct for each domain of individuals can actually be derived from them is still an unsolved question' (1928, p. 68).

This question Gödel answered in the affirmative in his Ph.D. thesis (Vienna, 1930), of which the paper under discussion is a rewritten version.

I shall not describe Gödel's proof. Perhaps no theorem in modern logic has been proved more often than Gödel's completeness theorem for the first-order predicate calculus. It stands at the focus of a complex of fundamental theorems, which different scholars have approached from various directions (e.g. Kleene, 1967, chapter VI).

The theorem says that, for each formula A of the first-order predicate calculus, either A is provable or its negation $\neg$ A is satisfiable in the domain $\{0, 1, 2, \ldots\}$ of the natural numbers. (So, if A is valid, A is provable.) Equivalently, each formula A either is 'refutable' (i.e. its negation is provable) or is satisfiable in $\{0, 1, 2, \ldots\}$.

A formula which is satisfiable in some nonempty domain cannot be refutable. Hence Gödel's completeness theorem includes the Löwenheim–Skolem theorem that, if a formula is satisfiable, it is satisfiable in a countably infinite domain. Thoralf Skolem in (1920), besides tightening up Leopold Löwenheim's argument (1915), extended the theorem to the case of a countable infinity of formulas. If they are simultaneously satisfiable in some nonempty domain, they are simultaneously satisfiable in the domain $\{0, 1, 2, \ldots\}$. Gödel treated this case too. The proof-theoretic alternative to simultaneous satisfiability clearly can involve only a finite number of formulas: if the infinitely many formulas are not simultaneously satisfiable, the conjunction of some finite subset of them is refutable. Hence, if each finite subset of a countably infinite set of formulas is simultaneously satisfiable (each in a respective domain), the whole set is simultaneously satisfiable. This is called 'compactness'.

If the formulas contain the symbol $=$ for equality with its usual interpretation, the foregoing summary must be modified to substitute 'finite or countably infinite domain' for 'countably infinite domain' or 'the domain $\{0, 1, 2, \ldots\}$'.

The completeness of logic, coupled with these satisfiability results, entails surprising incompleteness results for systems of axioms.

The one noticed first is the Skolem 'paradox' (1922). If axiomatic set theory as formulated in the first-order predicate calculus is consistent, the axioms must all be true for a countably infinite domain of sets, though it is a theorem of the theory that there are uncountable sets of sets.

Another such result is the existence of nonstandard models of arithmetic, i.e. of the theory of the natural numbers. Such models were first constructed by Skolem in (1933) and (1934). Not until Leon Henkin (1947), I believe, was it noticed that their existence is an almost immediate consequence of the

49

compactness part of Gödel's completeness theorem. To give the idea quickly, I shall use a function symbol ′ intended to express 'plus-one', besides the individual symbol 0 intended to express 'zero'. (The argument can be adapted to the predicate calculus without function symbols, as it is treated by Hilbert and Ackermann and by Gödel.) Let $A_0$, $A_1$, $A_2$, ... be a list of formulas proposed as axioms to characterize the natural numbers 0, 1, 2, .... We shall see that they cannot do so. We may assume the list to include the third and fourth Peano axioms (1889) for arithmetic (the first and second are provided by use of 0 and ′ in the symbolism). Now consider the expanded list $A_0$, $\neg\, i = 0$, $A_1$, $\neg\, i = 0'$, $A_2$, $\neg\, i = 0''$, ... where i is a new individual symbol. Each finite subset of these is simultaneously satisfied by the intended meanings of the symbols in $A_0$, $A_1$, $A_2$, ... and as value of i a natural number $n$ for which $\neg\, i = 0^{(n)}$ is not in the subset. So by compactness, all of them are simultaneously satisfied. With the usual interpretation of equality, and in view of the presence of the two Peano axioms, it is easy to see that the resulting model is isomorphic to one in which 0, 1, 2, ... are the values of 0, 0′, 0″, ... and (since $\neg\, i = 0$, $\neg\, i = 0'$, $\neg\, i = 0''$, ... are satisfied) the value of i is not a natural number — a nonstandard model of arithmetic.

What was new in Gödel's paper (1930)? On 14 August 1964, Gödel wrote to van Heijenoort (cf. the latter's source book (1967, p. 510)):

> As for Skolem, what he could justly claim, but apparently does not claim, is that, in his 1922 paper, he implicitly proved: 'Either A is provable or ~ A is satisfiable' ('provable' taken in an informal sense). However, since he did not clearly formulate this result (nor, apparently had made it clear to himself), it seems to have remained completely unknown, as follows from the fact that Hilbert and Ackermann in 1928 do not mention it in connection with their completeness problem.

Gödel solved this problem incisively; in the case that $\neg$ A is not satisfiable, Gödel established the provability of A in the strict sense of Hilbert's proof theory; and Gödel gave the supplemental argument to preserve the meaning of = in the case of satisfiability of $\neg$ A.

Now I turn to Gödel's famous paper on formally undecidable propositions (1931).

Partly in response to the appearance at the turn of the century of contradictions in set theory, David Hilbert had (since (1904)) proposed the formalist foundation of mathematics. His proposal called *first* for formalization of a suitable portion of the existing classical mathematics, including 'ideal' statements, which (unlike 'real' statements) do not have elementary intuitive meanings (1925, 1927). That is, this portion of mathematics should be embodied in a *formal system*. In a formal system, all the methods of constructing formulas to express mathematical propositions, and all the mathematical assumptions and principles of logic to be used in proving theorems, are to be governed by explicitly stated rules. The application of these rules shall only require working mechanically with the forms of the formulas, without taking into account their meanings.

*Second,* his programme called for a proof by 'safe' methods that the resulting formal system is consistent. 'Consistency' (now often called 'simple consistency') was to mean that no two configurations should exist which constitute proofs in the system of a formula A and of its negation $\neg$ A, respectively. Hilbert's consistency proofs were thus to be by a direct study of the structure of the formal system to be proved consistent. Therewith consistency problems for the first time became elementary in their statement, on the level of elementary number theory. They are problems in the new subject of 'proof theory' or 'metamathematics'. Previously, relative consistency proofs had been given from time to time by interpreting a new theory (received with skepticism) in an older theory (generally accepted), as for example in the demonstration that non-Euclidean geometry is consistent if Euclidean geometry is.[2]

It was certainly envisaged in Hilbert's Programme that the formalization of the portion of mathematics selected be '(simply) complete': each closed formula of the formal system should be either provable or refutable in it.

Gödel's paper (1931) shows that in a formal system obtained by combining Peano's axioms for the natural numbers with the logic of *Principia Mathematica* (Whitehead and Russell, 1910-13) the desired completeness is lacking.

Gödel's proof made use of what we now call 'arithmetization of metamathematics'. Alfred Tarski in his paper on *The Con-*

*cept of Truth in Formalized Languages* (1933) (in preparation since 1929) used the same idea, which he said later 'was developed far more completely and quite independently by Gödel' (the Tarski collection (1956, p. 278)). Gödel says,

> The formulas of a formal system ... in outward appearance are finite sequences of primitive signs ..., and it is easy to state with complete precision *which* sequences of primitive signs are meaningful formulas and which are not. Similarly, proofs, from a formal point of view, are nothing but finite sequences of formulas (with certain specifiable properties).

Thus, altogether, in studying a formal system we are studying a countable collection of objects. So these objects can be represented by natural numbers, either indices of the objects in an enumeration of them, or numbers assigned to them in some convenient way not using all the numbers — a 'Gödel numbering' as we now say. I shall assume that a fixed Gödel numbering has been adopted.

Now various predicates of natural numbers can be formulated which, via the Gödel numbering, say things about the formal system. Can these predicates be expressed *in* the system? Generally, yes.

Indeed, Gödel investigated a series of 46 functions and predicates. He systematized his work by first describing a class of number-theoretic functions (i.e. functions $\varphi$ $(x_1, \ldots, x_n)$ where $x_1, \ldots, x_n$ range over the natural numbers and the values are natural numbers) which he called 'recursive', and which had already been used in foundational investigations, e.g. by Dedekind (1888), Skolem (1923), Hilbert (1925, 1927) and Ackermann (1928). Gödel gave a precise definition of this class of functions, which has become standard. In a paper (1936), I changed the name to 'primitive recursive functions', which has become standard. 'Primitive recursive' predicates are predicates $P(x_1, \ldots, x_n)$ such that $P(x_1, \ldots, x_n) \leftrightarrow \varphi(x_1, \ldots, x_n) = 0$ for a primitive recursive function $\varphi$.

Gödel showed that the first 45 of his functions and predicates are primitive recursive. (His predicate numbered 46 is '$x$ is the Gödel number of a provable formula', and predicate 45 is '$x$ is the Gödel number of a formula, and $y$ is the Gödel number of a proof of that formula'. Given $x$ and $y$, it can be determined whether the latter is true or false by recapturing from $x$ and $y$

the alleged formula and proof and *examining their parts.* To determine whether predicate 46 is true for a given $x$, one faces an open-ended search for a $y$ such that $x$ and $y$ make predicate 45 true.)

Furthermore, Gödel showed that, to each primitive recursive predicate $P(x_1, \ldots, x_n)$, there is a formula $P(x_1, \ldots x_n)$, with corresponding formal variables $x_1, \ldots x_n$ (and no others) free in it, such that the following holds, for each $n$ natural numbers $x_1$, $\ldots, x_n$:[3] Let $P(x_1, \ldots, x_n)$ be the formula which comes from $P(x_1, \ldots, x_n)$ by substituting, for (the free occurrences of) the variables $x_1, \ldots, x_n$, the numerals $x_1, \ldots, x_n$ (formal expressions of the system) expressing the numbers $x_1, \ldots, x_n$. If $P(x_1, \ldots, x_n)$ is a true proposition about the numbers $x_1, \ldots, x_n$, then the formula $P(x_1, \ldots, x_n)$ is provable. If $P(x_1, \ldots, x_n)$ is false, then $P(x_1, \ldots, x_n)$ is refutable. In terminology introduced by me in (1952), the formula $P(x_1, \ldots, x_n)$ 'numeralwise expresses' the predicate $P(x_1, \ldots, x_n)$, which (when such a formula exists) is said to be 'numeralwise expressible', in the formal system.

As one would expect, the formula $P(x_1, \ldots, x_n)$ constructed to numeralwise express $P(x_1, \ldots, x_n)$ also expresses $P(x_1, \ldots, x_n)$ under the usual interpretation of the symbolism.

Gödel showed, moreover, that a formula expressing (and numeralwise expressing) a primitive recursive predicate can always be found in the language having only number variables, 0 and 1, $+$ and $\cdot$ (addition and multiplication), and the logical symbolism of the first-order predicate calculus. Predicates so expressible Gödel called 'arithmetical'.

Now I can represent Gödel's reasoning faithfully, in the time available, by using some notations of my own.

To fix our ideas, let us consider formulas which contain free just the one particular variable v. For any number $g$ which *is* the Gödel number of such a formula, I write that formula as $A_g(v)$.

Let $A(v, x)$ be the predicate which says the following: $v$ is the Gödel number of a formula $A_v(v)$ containing free just the variable v, and $x$ is the Gödel number of a proof of the formula $A_v(\textit{v})$ (which comes from $A_v(v)$ by substituting for v the numeral $\textit{v}$ expressing $v$).

No person ignorant of the Gödel numbering would think of the predicate $A(v, x)$; but (with the Gödel numbering adopted) it is precisely definable in number-theoretic terms. Working through Gödel's 45 primitive recursive functions and predicates, $A(v, x)$ is primitive recursive.

After choosing a formula A(v, x) that expresses and numeralwise expresses $A(v, x)$, consider the formula $\forall x \neg$ A(v,x), which says 'for all $x$, not $A(v, x)$'. This formula has the one free variable v, and of course it has a Gödel number $p$, so it is the formula $A_p(v)$.

Finally, consider the formula $A_p(p)$, which is $\forall x \neg A (p, x)$.

What does $A_p$ $(p)$ say under the meaning of the symbolism via the Gödel numbering? We already know that $p$ is the Gödel number of a formula $A_p(v)$ containing free just the variable v. So, for each $x$, $A(p, x)$ is equivalent to '$x$ is the Gödel number of a proof of $A_p(p)$'. So $A_p(p)$, which is $\forall x \neg A (p, x)$, says that, for all $x$, $x$ is not the Gödel number of a proof of $A_p(p)$. In brief, $A_p(p)$ says, 'I am unprovable'.

We know that the sentence which says 'I am false' is paradoxical. For, if it is true, then by what it says it is false; and if it is false, then by what it says it is true.

Gödel's sentence 'I am unprovable' is not paradoxical. We escape paradox because (whatever Hilbert may have hoped) there is no a priori reason why every true sentence must be provable.

The sentence $A_p(p)$, which says 'I am unprovable', is simply unprovable and true.

This conclusion is inescapable, *if* in the formal system only true sentences are provable. For, if $A_p(p)$ were provable, then by what it says it would be false, contradicting our supposition that only true sentences are provable. So $A_p(p)$ is unprovable, and then by what it says it is true.

Under the same hypothesis, $\neg A_p(p)$ is also unprovable, since it is false.

Thus, if only true sentences are provable, $A_p(p)$ is 'formally undecidable' in the system, i.e. it is neither provable nor refutable.

This is an 'informal' version of Gödel's (first) incompleteness theorem. Rather than assume simply that only true sentences are provable, a formalist or metamathematician would prefer to substitute consistency properties.

To establish that $A_p(p)$ is unprovable, it suffices to assume that the system is simply consistent. For, if $A_p(p)$ were provable, there would be a proof of it, which would have a Gödel number $x$. For this $x$, $A(p, x)$ would be true; so, because A(v, x) numeralwise expresses $A(v, x)$, the formula A(p, x) would be provable, and thence so would be $\exists x A(p, x)$ and $\neg \forall x$

$\neg$ A $(p, x)$, which is $\neg$ A$_p(p)$. So both A$_p(p)$ and $\neg$ A$_p(p)$ would be provable, which would constitute a simple inconsistency.

To establish that $\neg$ A$_p(p)$ is unprovable, Gödel used what he called '$\omega$-consistency': for no formula A(x) are all of A(0), A(1), A(2), ... and $\neg$ $\forall$xA(x) provable. This is a property called for under the interpretation of the system when x is a variable interpreted to range over the natural numbers. The unprovability of $\neg$ A$_p(p)$ follows from $\omega$-consistency, thus: $\omega$-consistency implies simple consistency; and simple consistency implies the unprovability of A$_p(p)$; thence each of 0, 1, 2, ... is not the Gödel number of a proof of A$_p(p)$; thus $A(p,0)$, $A(p, 1)$, $A(p, 2)$, ... are false; so $\neg$ A(p, 0), $\neg$ A(p, 1), $\neg$ A(p, 2), ... are provable; so if $\neg$ A$_p(p)$, which is $\neg$ $\forall$x $\neg$ A(p, x), were provable, the system would be $\omega$-inconsistent.

Thus, if the system is $\omega$-consistent, A$_p(p)$ is formally undecidable (Gödel's (first) incompleteness theorem, metamathematical version).

Incidentally, the formally undecidable proposition is arithmetical.

If the formal system in question is simply consistent, then the addition of the formula $\neg$ A$_p(p)$ as a new axiom would produce a simply consistent but $\omega$-inconsistent system. For, if the enlarged system were simply inconsistent, then in the unenlarged system $\neg$ $\neg$ A$_p(p)$ would be provable, whence A$_p(p)$ would be provable; but A$_p(p)$ is not provable if the system is simply consistent. This shows that just the simple consistency of a system (which is what Hilbert had in mind to prove) would not guarantee its correctness under the interpretation when there are variables interpreted as ranging over the natural numbers.

Barkley Rosser in (1936) replaced Gödel's A$_p(p)$ by a slightly more complicated formula A$_p(q)$ whose formal undecidability follows from just simple consistency.

Let us return to the first part of Gödel's formal undecidability result. No sooner had we concluded that A$_p(p)$ is unprovable than we knew it is true.

What is lacking *in the system* that a formula we can recognize to be true is not provable in it?

Consider our result more precisely. We concluded that A$_p(p)$ is true *under the hypothesis* that in the system provable-implies-true, or in the metamathematical version that the system is simply consistent. All the reasoning to show that $A_p(p)$ is unprovable and hence true, when carried out in full detail, is of the character of elementary number theory, except that we put that hypothesis into it. In brief, we have shown by elementary number-theoretic reasoning that:

(1) If the formal system is simply consistent, then $A_p(p)$ is true.

The proposition that the formal system is simply consistent can be expressed via the Gödel numbering by a formula of the system. One way is to observe that $0 = 0$ is provable. Let $r$ be the Gödel number of $\neg 0 = 0$, and let the formula say that, for all $x$, $x$ is not the Gödel number of a proof of the formula with Gödel number $r$. (From any contradiction in the system; $\neg 0 = 0$ would follow.) Call such a formula 'Consis'.

The observation that we have established (1) by elementary number-theoretic reasoning suggests that the formula expressing (1) is provable in the system; i.e. that:

(2)         The formula Consis $\supset A_p(p)$ is provable.

Gödel planned to establish (2) in detail in a second part of the paper; but it seems no one doubted it. Hilbert and Bernays in Volume 2 of *Grundlagen der Mathematik* (1939) carried out this exercise in a similar system. Let us accept (2). Then Consis cannot be provable in the system, *if* the system *is* simply consistent (Gödel's second incompleteness theorem). For, if Consis were provable, then from it and (2) by modus ponens, $A_p(p)$ would be provable; and we have seen that $A_p(p)$ is not provable if the system is simply consistent.

This is the second way in which Gödel's paper (1931) showed that Hilbert's formalist programme cannot be carried out in any simple way. Hilbert hoped, after formalizing the core of classical mathematics, to show the resulting system consistent by what he called in German 'finit' means, which in my book (1952) I translated as 'finitary'. The simplest way to describe finitary methods is to say that they are ones that do not involve the conception of the completed infinite. But now we find that not even all the means formalized in the system will suffice for showing its consistency if it is consistent.

This does not rule out absolutely the possibility of a finitary consistency proof for a formalism embodying at least elementary number theory. Rather, as Gödel observed, it is conceivable that some method exists which is not included in the formalism but can be construed as finitary and which would suffice for a consistency proof.

Indeed, just this has happened. Nearly a decade of unsuccessful attempts at consistency proofs for systems encompassing elementary number theory had preceded Gödel's paper (1931). In September of 1935, when I was about to spend a year at the Institute for Advanced Study, Hermann Weyl put in my hands a manuscript by Gerhard Gentzen which he wished me to read. I was prevented from doing so by a telegram offering me a teaching position at the University of Wisconsin. This manuscript contained a consistency proof for elementary number theory, of which a revised version was published in (1936) (cf. Paul Bernays, 1968). It proffered, as a nonelementary but finitary method, induction over the segment of Cantor's ordinal numbers up to the first $\varepsilon$-number $\varepsilon_0$. Other proofs, also using $\varepsilon_0$-induction, appeared subsequently (cf. Kleene, 1967, §44).

It has been little noticed that the incompleteness of first-order formal systems for number theory (Gödel, 1931) and the completeness of the first-order logical calculus (Gödel, 1930) taken together (without using compactness as above) imply the existence of nonstandard models of effectively given axiom systems of arithmetic. Briefly, a number-theoretic formal system consists of logic plus number-theoretic axioms. If the logic is complete but the whole is incomplete, the number-theoretic axioms must be incomplete. I have found only a two-line allusion to this in Gödel's *Zentralblatt* review (1934a, p. 194) of Skolem (1933), and three lines in Henkin (1950, p. 91), prior to my independent discussion of it in (1952, p. 430).

Might Gödel's results apply only to his particular system and a class of related systems?

Gödel (1931) went to some pains to indicate the generality of his results. Thus he observed that they would hold good for the formal systems having the Zermelo–Fraenkel axiom system, or that of von Neumann, for set theory, or the Peano axioms for the natural numbers and the schema of (primitive) recursion, besides the rules of (first-order) logic.

In a note *On completeness and consistency* (1931-2), he starts with the last-mentioned system, and considers successive enlargements by adding higher-type variables (for classes of numbers, classes of classes of numbers, etc.). He thus obtains a sequence of formal systems (continuable into the transfinite) in which some undecidable propositions of earlier systems become decidable, while new undecidable propositions are constructible by the same procedure. In a paper (1936) *On the length of proofs*, he states that in systems with higher types of variables, and under the appropriate consistency assumptions, not only do some previously unprovable sentences become provable, but infinitely many of the previously available proofs can be very greatly shortened.

The ideas by which the fullest generality of Gödel's results were made definitive were in 1931 only about to be conceived.

In the spring semester of 1933-34, Gödel lectured at the Institute for Advanced Study *On undecidable propositions of formal mathematical systems*. At the suggestion of Oswald Veblen, Rosser and I produced a set of notes (Gödel (1934)). The final section is entitled 'General recursive functions'. Gödel, restricting in two respects a definition suggested to him (in a letter) by Jacques Herbrand, gave a definition of a class of functions extending very greatly the primitive recursive functions (cf. van Heijenoort, 1967, p. 619). This gave rise to the question whether the general recursive functions do not encompass all number-theoretic functions for which algorithms exist, or which (to use a phrase of Alonzo Church) are 'effectively calculable'. For predicates, we say 'effectively decidable'. The same question had previously arisen for another class of functions, called the 'λ-definable functions'. The notion of the λ-definable functions was due jointly to Church and the present writer, whose investigations of them, begun early in 1932, were included in his Ph.D. thesis, 1933, and published in (1935) (cf. Church, 1936, footnotes 3 and 18). There followed in Church (1936) what the present writer in (1952), if not quite in (1943) called 'Church's thesis'. This identifies the effectively calculable functions with the general recursive functions, or equivalently (Church, 1936; Kleene, 1936a) with the λ-definable functions.

This enabled the writer to give some generalized versions of Gödel's incompleteness theorem (the first in (1936)). In these generalized versions, the effectiveness requirements that a formal system must possess for Hilbert's purpose are identified, in view of Church's thesis, with requirements of general recursiveness (or an equivalent). In these versions the formally undecidable propositions, which every formal system in question must possess, are values of an arithmetical predicate $P(x)$ picked in advance. Thus no collection of correct logical principles and mathematical axioms that one could ever be able to describe effectively will suffice for deciding the truth or falsity of $P(x)$ for every $x$.

The idea of the version in my paper (1943) (reproduced in (1952, §60)) is this. Suppose we had a formal system which is complete and correct for the theory of a predicate $P(x)$. Then, for each $x$, the proposition $P(x)$ would be true exactly if $(Ey)R(x, y)$, where '$(Ey)$' says 'there exists a $y$', and $R(x, y)$ says '$y$ is the Gödel number of a proof of the formula expressing $P(x)$ for that $x$'. By the effectiveness requirements for a formal system taken with Church's thesis, $R(x, y)$ should be general recursive. (In (1943), I called Church's thesis 'Thesis I' and this application of it to formal systems 'Thesis II'.) In brief, the existence of a complete and correct formalization of the theory of a predicate $P(x)$ would impose on $P(x)$ that it be expressible in the form $(Ey)R(x, y)$ with $R(x, y)$ general recursive. But there are predicates $P(x)$ which aren't expressible in that form. Such predicates can be given of the form $(y)S(x, y)$ where '$(y)$' says 'for all $y$' and $S(x, y)$ is primitive recursive. (This is proved using $\bar{T}_1(x, x, y)$ as the $S(x, y)$ in Kleene (1943) and (1952), and, even in (1936, Theorems III and XVI with footnote 22) and for constructiveness the addendum in Davis (1965, p. 253); and in Kleene (1967, pp. 243, 270) using $\bar{T}(x, x, y)$ as the $S(x, y)$.)

Subsequent to the Gödel notes (1934) and the papers of Church and me (1936) there appeared the important paper of Alan Mathison Turing (1936-7). This reinforced Church's thesis by giving a new notion, that of a 'Turing computable function', under which the thesis becomes especially convincing and which was subsequently proved equivalent to 'general recursive function' and to 'λ-definable function'. Still other equivalents have appeared, as by Gödel (1936), Emil L. Post (1936, 1943), A.A. Markov (1951), Kleene (1959) and J.C. Sheperdson and H.E. Sturgis (1963).

Gödel lent his assent to the generalized Gödel theorems in the words of the following 'Note added 28 August 1963' to the translation of his paper (1931) in van Heijenoort (1967):

In consequence of later advances, in particular of the fact that due to A.M. Turing's work a precise and unquestionably adequate definition of the general notion of formal system can now be given, a completely general version ... is now possible. That is, it can be proved rigorously that in *every* consistent formal system that contains a certain amount of finitary number theory there exist undecidable arithmetic propositions and that, moreover, the consistency of any such system cannot be proved in the system.

I have dwelt on developments growing out of Gödel's work to illustrate how seminal his ideas and results have been.

I will conclude my discussion of his discovery of formally undecidable propositions by quoting from what John von Neumann said on 14 March 1951, on the occasion of an award of an Einstein medal to Gödel: 'Kurt Gödel's achievement in modern logic is singular and monumental — indeed it is more than a monument, it is a landmark which will remain visible far in space and time' (*New York Times*, 15 March 1951, p. 31).

Before passing to the third of Gödel's major contributions that I am discussing in detail, I will mention a number of observations that he contributed to the study of intuitionistic systems. In (1932) he showed that the intuitionistic propositional calculus cannot be treated on the basis of finitely many truth values. In (1932-3) he showed that it is possible to paraphrase every theorem in classical number theory by one in intuitionistic number theory; specifically, $\exists x A(x)$ is paraphrased as $\neg \forall x \neg A(x)$, $A \vee B$ as $\neg (\neg A \& \neg B)$, and $A \supset B$ as $\neg (A \& \neg B)$. (Later Bernays, and Gentzen (1936, p. 532), found this result independently without paraphrasing $A \supset B$.) Thus Gödel wrote, 'The theorem shows that the intuitionistic arithmetic and number theory is only apparently narrower than the classical; in fact [it] includes the entire classical [number theory], merely with a somewhat differing interpretation.' Arend Heyting in (1934, p. 18) added, 'However for the intuitionists this interpretation is the essential thing.' A paper of Gödel (1958) *On a previously unused extension of the finitist standpoint* has been playing a role in current investigations of models of intuitionistic mathematics.

I am not qualified to give an account of Gödel's work in relativity theory and in philosophy.

The third major contribution of Gödel which I shall describe is his use of an inner model of set theory to establish the consistency of the axiom of choice and of the generalized continuum hypothesis with the (other) axioms of set theory (1938, 1939, 1940).

Georg Cantor's set theory, initiated in his paper (1874), develops a theory of (finite and) infinite magnitudes.

For two sets $M$ and $N$, let their *magnitudes* (Mächtigkeiten) or *cardinal numbers* be card $M$ and card $N$. Then card $M =$ card $N$ under Cantor's definition, iff $M$ can be put into one-to-one correspondence with $N$; card $M <$ card $N$ (or card $N >$ card $M$), iff $M$ can be put into one-to-one correspondence with a part or *subset* of $N$ (i.e. any set whose members are members of $N$), but not vice versa.

A set $M$ (or its cardinal card $M$) is *finite*, iff $M$ can be put into one-to-one correspondence with an initial segment $\{0, 1, \ldots, n - 1\}$ of the natural numbers (then card $M = n$); *infinite*, otherwise.

It at once became of interest in Cantor's theory to relate to one another the infinite cardinals first encountered.

The cardinal number $\aleph_0$ of the set $\omega$ of the natural numbers is an infinite cardinal, and no cardinal lies properly between it and the finite cardinals, i.e. for no $M$ is $0, 1, 2, \ldots <$ card $M <$ $\aleph_0$.

The real numbers, or the points of the real line or linear continuum, have a greater cardinal than the natural numbers.

Let $\mathscr{P}(M)$ be the set of the subsets of $M$ (including $M$ itself and the empty set), called the 'power set of $M$'. Let its cardinal card $\mathscr{P}(M)$ also be written $2^{\operatorname{card} M}$ (which is consistent with ordinary arithmetic in the case that $M$ is finite). According to Cantor's theorem (1890–1), for each set $M$, $2^{\operatorname{card} M} >$ card $M$; in particular, $2^{\aleph_0} > \aleph_0$.

The cardinal $2^{\aleph_0}$ is the same as the cardinal of the real numbers; i.e. there is a one-to-one correspondence between $\mathscr{P}(\omega)$ and the real numbers.

One form of the *continuum hypothesis* is that each subset of the continuum has either the cardinal $\aleph_0$ of the natural numbers or the cardinal $2^{\aleph_0}$ of the whole continuum; or equivalently, that for no $M$ is $\aleph_0 <$ card $M < 2^{\aleph_0}$.

Another form arises from Cantor's theory of (finite and) transfinite ordinals and their cardinals. *Ordinal numbers* are associated with *well-ordered sets*, i.e. sets which are linearly ordered so that each nonempty subset has a least element. Let two such sets $A$ and $B$ have the ordinals $\alpha$ and $\beta$. Then $\alpha = \beta$, iff $A$ and $B$ can be put into one-to-one correspondence preserving the ordering; and $\alpha < \beta$ (and $\beta > \alpha$), iff $A$ can be put into a one-to-one order-preserving correspondence with an initial segment of $B$, i.e. with the elements of $B$ preceding a given element of $B$. For each two ordinals $\alpha$ and $\beta$, either $\alpha < \beta$ or $\alpha = \beta$ or $\alpha > \beta$. The ordinals themselves form a well-ordered collection.

It has become standard, following von Neumann (1923), to identify each ordinal with the set of the ordinals $<$ it, so that $<$ becomes the membership relation $\in$ between ordinals considered as sets. This makes the first infinite ordinal $\omega$ ($= \omega_0$) the set of the natural numbers (used as ordinal numbers), as we have been saying.

Now it can be proved that the set of the ordinals which can be put into one-to-one correspondence with $\omega$ (i.e. which have the cardinal $\aleph_0$), called Cantor's 'second number class', has a cardinal $\aleph_1$ greater than the cardinal $\aleph_0$ of $\omega$; and that, for no $M$, is $\aleph_0 < \text{card } M < \aleph_1$.

A second form of the *continuum hypothesis* is that $2^{\aleph_0} = \aleph_1$.

$\aleph_1$ is the cardinal of the first ordinal $\omega_1$ of Cantor's 'third number class'. Continuing in this fashion, Cantor obtained a succession of transfinite cardinals (the 'alephs') $\aleph_0, \aleph_1, \aleph_2, \ldots,$ $\aleph_\alpha, \ldots$ increasing with $\alpha$ where $\alpha$ ranges over all ordinals, which are all the cardinals of well-ordered infinite sets, and which are the cardinals of the first ordinals $\omega_0, \omega_1, \omega_2, \ldots, \omega_\alpha,$ $\ldots$ of respective number classes, the 2nd, 3rd, 4th, $\ldots, (2 + \alpha)$ $-$ th, $\ldots$ . If card $M < \aleph_\alpha$, then card $M$ must be an aleph.

The *generalized continuum hypothesis* is that, for all $\alpha$, $2^{\aleph_\alpha} = \aleph_{\alpha + 1}$.

At this stage of the theory the unhappy possibility remained open that pairs of cardinals might exist incomparable with each other.

Then Ernst Zermelo in (1904) and (1908) proved that, if one assumes his *axiom of choice*, then every set can be well-ordered, so (as Cantor had believed but not proved) every infinite cardinal is an aleph, and every two cardinals are comparable.

The axiom of choice in one form states that, given a set $S$, there exists a function $f$ such that for each nonempty subset $S_0$ of $S$, $f(S_0) \in S_0$. This function $f$ is a set, i.e. a set of ordered pairs $(T, U)$ such that for each set $T$ there is at most one $U$ for which $(T, U)$ is in the set. There are other forms of the axiom of choice.

Just sixty years elapsed from Cantor's enunciation of the continuum hypothesis as a conjecture in (1878, p. 257) to the first really significant result on it by Gödel in (1938). The continuum problem was the first in Hilbert's famous list of mathematical problems (1900).

Gödel's accomplishment is in the context of axiomatic set theory.

From Cesare Burali-Forti's paper (1897), it became known that contradictions arise in set theory, pursued naively from Cantor's definition of a set as 'any collection $M$ of definite well-distinguished objects $m$ of our perception of our thought (which are called the 'elements' of $M$) into a whole' (1895-7, the first sentence). Thus, the Burali-Forti paradox arises by construing the ordinals as a set under this definition, which set is well-ordered and so has an ordinal, which must be greater than all ordinals.

Hence, with the aim of making set theory consistent, systems of axioms have been formulated to say just what sets shall exist and just what shall be assumed about them. This is a less drastic response to the paradoxes than Hilbert's formalism.

The first axiomatization of set theory was by Zermelo in (1908a). Subsequent writers were troubled by the impreciseness of the idea of 'definite property' used in Zermelo's *axiom of separation*. This axiom asserts that the elements $x$ of a given set $y$ which have a definite property shall constitute a set. Abraham A. Fraenkel proposed a remedy in (1922), Skolem in (1922), and Weyl even earlier (1910, 1918), dealt with the problem somewhat differently.

For simplicity, I shall talk about a system of axioms usually called Zermelo–Fraenkel set theory, but which is closer to Skolem and Weyl on the point in question. The axioms will be expressed in the symbolism of the first-order predicate calculus with just two predicate symbols, $\in$ and $=$, where '$x \in y$' says that $x$ is a *member* or *element* of $y$ or *belongs* to $y$. I shall denote by 'ZF' this system without the axiom of choice AC, and by 'ZFC' the same with AC. In ZF, Zermelo's axiom of separation is replaced by the *axiom schema of comprehension*, which says that the elements $x$ of a given set $y$ which satisfy a formula $\varphi(x)$ of the system (or with parameters, $\varphi(x, z_1, \ldots, z_n)$) constitute a set. ZF has two primitive mathematical concepts (besides the logical ones, including $=$): *set*, which enters implicitly as the range of the variables, and *membership* $\in$.

63

Gödel used explicitly *set* $\mathfrak{M}$, *class* $\mathfrak{Cls}$ and *membership* $\in$.

In the account I am giving now, classes will enter only as collections of the $x$'s for which some formula $\varphi(x)$, or $\varphi$ $(x, z_1, \ldots, z_n)$, is true, and designations of classes will enter only in abbreviations for formulas. Only sets can be members of sets; classes are not necessarily sets.

For example, the class of the ordinals is definable by a formula of ZF, i. e. as the sets $\alpha$ satisfying that formula; and, in lieu of writing out the formula, I shall use '$\alpha \in$ On' to stand for it. The class of the ordinals is not a set; if it were, we would have the Burali-Forti paradox.

What Gödel did was to define within the theory provided by the axioms of ZF a class $L$ of sets called the *constructible sets*.

I shall summarize the treatment Gödel gave in lectures in the fall term of 1938-39, available through notes by George W. Brown (Gödel (1940)).

A first crucial step concerns the axiom schema of comprehension. This is a bundle of a countable infinity of axioms, according to the possible choices of the formula $\varphi$ $(x)$ or $\varphi$ $(x, z_1, \ldots, z_n)$. Gödel recognized that these infinitely many axioms can be replaced by a finite number of them. Indeed, iteration of eight operations, $\mathscr{F}_1, \ldots, \mathscr{F}_8$ suffices to give the full effect of the axiom schema of comprehension. Gödel took all of these to be binary operations; i.e. each operates on two sets $X$, $Y$ to produce a third $Z$.

Gödel defined a function $F(\alpha)$ by transfinite recursion over the ordinals. This $F$ is only a class; i.e. the relation $Z = F(\alpha)$ is expressed by a formula of ZF, but the collection of all ordered pairs $(\alpha, F(\alpha))$ for $\alpha \in$ On is not a set.

To define $F(\alpha)$, Gödel first correlated to each ordinal $\alpha$ a triple $(i, \beta, \gamma)$. As $\alpha$ ranges over On, we get exactly once each possible triple for $i = 0, \ldots, 8$, for $\beta \in$ On and for $\gamma \in$ On. We always have $\beta, \gamma < \alpha$, except that $\beta = \gamma = 0$ for $\alpha = 0$; and for given $\beta$, $\gamma$ the values $0, \ldots, 8$ of $i$ are taken consecutively.

Now if $\alpha$ corresponds to $(i, \beta, \gamma)$ with $1 \le i \le 8$ (so $\beta$, $\gamma < \alpha$), $F(\alpha) = \mathscr{F}_i (F(\beta), F(\gamma))$. If $\alpha$ corresponds to $(0, \beta, \gamma)$, then $F(\alpha)$ is the set of all $F(\xi)$ for $\xi < \alpha$. (So $F(0)$ is the empty set.)

Finally, the class $L$ of the constructible sets is the collection of all $F(\alpha)$ for $\alpha \in$ On.

Now it turns out that all the axioms of ZF when relativized to $L$ become theorems of ZF. Briefly, 'relativized to $L$' means that the variables are restricted to range just over the sets belonging to $L$. Thus, for each axiom $A$, its relativization $A^L$ is a theorem of ZF; i.e. $A^L$ is deducible by logic from the (unrelativized) axioms of ZF.

Furthermore, the same is true of the formula $V = L$ where $V$ is the class of all sets; i.e. $[V = L]^L$ is a theorem of ZF. Here we can express $V = L$ in ZF as $(x)(Ea)[a \in \text{On } \& \ x = F(a)]$.

It follows that, if ZF is consistent, so is ZF $+ \ V = L$. For, suppose a contradiction, say $0 = 0 \ \& \ \neg 0 = 0$, or simply $\neg 0 = 0$ (since $0 = 0$ is provable), exists in the system ZF $+ \ V = L$. This means that there is a logical deduction of $\neg 0 = 0$ from the axioms of ZF and the formula $V = L$.[4] Say that the axioms of ZF used in this deduction are $A_1, \ldots, A_m$. Then there would be a parallel deduction of $[\neg 0 = 0]^L$, which is $\neg 0 = 0$, from $A_1^L$, $\ldots, A_m^L, [V = L]^L$. Combining this deduction with the proofs of $A_1^L, \ldots, A_m^L, [V = L]^L$ as theorems of ZF, we would obtain a proof of $\neg 0 = 0$ in ZF itself.

Now, finally, the axiom of choice AC and the generalized continuum hypothesis GCH ($2^{\aleph_a} = \aleph_{a+1}$) are theorems in the system ZF $+ \ V = L$.

This is easily seen for AC, since AC would follow if we had the existence of a well-ordering of the class $V$ of all sets. But the generation of $L$ as the range of $F(a)$ when $a$ ranges over On induces a well-ordering of $L$. When $V = L$ is assumed, this is a well-ordering of $V$.

For GCH, Gödel shows that, when $\omega_a$ is the first ordinal in Cantor's $(2 + a)$th number class ( $=$ the least ordinal with the cardinal $\aleph_a$), then the set of values of $F(\xi)$ for $\xi < \omega_a$ (write it $F``\omega_a$) has the cardinal $\aleph_a$. Furthermore, he shows that each subset of $F``\omega_a$ that belongs to $L$ will arise as value of $F(\xi)$ for some $\xi < \omega_{a+1}$. Thus $2^{\aleph_a} = \text{card } \mathscr{P}(F``\omega_a) \leq \text{card } F``\omega_{a+1} = \aleph_{a+1}$. But by Cantor's theorem, $\aleph_a < 2^{\aleph_a}$. So $2^{\aleph_a} = \aleph_{a+1}$.

In summary, AC and GCH are theorems of ZF $+ \ V = L$, and ZF $+ \ V = L$ is consistent if ZF is. So, if ZF is consistent, ZF $+$ AC $+$ GCH is consistent (being a subsystem of the consistent system ZF $+ \ V = L$).

It was remarked by Bernays in (1940) that Gödel's construction of $L$ can be regarded as a modification of an uncompleted project of Hilbert (1925) for a proof of the continuum hypothesis. (Since Cohen (1963-4), we have known that Hilbert's project could not have succeeded with only the axioms of ZFC.) Gödel wrote van Heijenoort on 8 July 1965 (van

65

Heijenoort (1967, p. 368)) acknowledging a 'remote analogy' and continuing:

> There is, however, this great difference that Hilbert considers only strictly constructive definitions and, moreover, transfinite iterations of the defining operations only up to constructive ordinals, while I admit, not only quantifiers in the definitions, but also iterations of the defining operations up to *any* ordinal number, no matter whether or not it can be defined. The term 'constructible set', in my proof, is justified only in a very weak sense, and, in particular, only in the sense of 'relative to ordinal numbers', where the latter are subject to no conditions of constructivity. It was exactly by viewing the situation from this highly transfinite, set-theoretic point of view that in my approach the difficulties were overcome and a *relative* finitary consistency proof was obtained.

Of course, Gödel in (1938) did not 'solve' Cantor's continuum problem in the way Cantor surely had in mind. Cantor in (1878), and in (1884) when he mistakenly thought that he had proved his continuum hypothesis (cf. Church, 1966), was not thinking of its status relative to a list of axioms. After choosing an axiomatization, say ZFC, there are three possibilities respecting, say, the simple continuum hypothesis $2^{\aleph_0} = \aleph_1$. First, it is provable from the axioms. Second, it is refutable from the axioms. Third, it is undecidable from the axioms. Gödel's work eliminated the second possibility.

In an expository paper, *What is Cantor's continuum problem?* (1947), Gödel expressed the views that 'the axiom of choice ... is, in the present state of knowledge, exactly as well founded as the system of the other axioms' (note 2) but 'one may on good reason suspect that the role of the continuum problem in set theory will be this, that it will finally lead to the discovery of new axioms which will make it possible to disprove Cantor's conjecture' (p. 524).

Paul J. Cohen in (1963-4), using a quite different model than Gödel's *L*, showed that the negation of the continuum hypothesis is consistent with ZFC, thus eliminating the first possibility. So $2^{\aleph_0} = \aleph_1$ is undecidable from the axioms of ZFC.

Gödel's result (1938), joined to Cohen's (1963-4), set the stage for a whole new era in the theory of sets in which a host of problems of the consistency or independence of various conjectures in set theory relative to this or that set of axioms are being investigated by constructing models. I cannot imagine that Cantor ever dreamed that the subject he brought into the mathematical world just over a century ago would take on this aspect. What will it be like in another hundred years?

We can be certain of one thing: Gödel's work here, as well as in the demonstration of undecidability phenomena in formal number theory and in a definitive treatment of the completeness of first-order logic, will have remained a landmark whose visibility from afar will have been undiminished by time.

## NOTES

1. A Survey Lecture delivered to the Association for Symbolic Logic at New York, N.Y. on 29 December 1975.
2. By formalizing both theories à la Hilbert, one can thence obtain consistency proofs on the level of elementary number theory for the one theory (as embodied in a formal system) relative to the other (similar embodied).
3. I am using *italic* letters to name predicates and (variable) natural numbers of your and my language (the metalanguage), Roman letters to name formulas and variables of the formal system, and **bold face italic** letters to name numerals.
4. Gödel says (1940, p. 1), 'Although the definitions and theorems are mostly stated in logistic symbols, the theory developed is not to be considered as a formal system but as an axiomatic theory in which the meaning and the properties of the logical symbols are presupposed to be known. However to everyone familiar with mathematical logic it will be clear that the proofs could be formalized, using only the rules of Hilbert's "Engere Funktionenkalkül" [with equality].' It is by such formalization that a finitary consistency proof is afforded (cf. Note 2).

## REFERENCES

Ackermann, Wilhelm (1928) Zum Hilbertschen Aufbau der reellen Zahlen. *Mathematische Annalen* 99, 118-33.
Bernays, Paul (1940) Review of Gödel (1939). *Journal of Symbolic Logic*, 5, 117-18.
—— (1968) On the original Gentzen consistency proof for number theory. *Intuitionism and proof theory, Proceedings of the Summer Conference at Buffalo, N.Y.*, 1968, North-Holland, Amsterdam and London, 1970, 409-17.
Burali-Forti, Cesare (1897) Una questione sui numeri transfiniti. *Rendiconti del Circolo Matematico di Palermo*, 11, 154-64.

Cantor, Georg (1874) Über eine Eigenschaft des Inbegriffes aller reellen algebraischen Zahlen. *Journal für die reine und angewandte Mathematik*, 77, 258-62.

—— (1878) Ein Beitrag zur Mannigfaltigkeitslehre, Ibid., 84, 242-58.

—— (1884) Über unendliche, lineare Punktmannigfaltigkeiten, *Mathematische Annalen*, 23, 453-88.

—— (1890-1) Über eine elementare Frage der Mannigfaltigkeitslehre. *Jahresbericht der Deutschen Mathematiker-Vereinigung*, 1, 75-8.

—— (1895-7) Beiträge zur Begründung der transfiniten Mengenlehre, *Mathematische Annalen*, 46 (1895), 481-512 and 49 (1897), 207-46.

Church, Alonzo (1936) An unsolvable problem of elementary number theory. *American Journal of Mathematics*, 58, 345-63.

—— (1944) *Introduction to mathematical logic. Part I*. Annals of Mathematics Studies, no. 13. Princeton University Press, Princeton, N.J.

—— (1956) *Introduction to mathematical logic, Vol. 1*. Princeton University Press, Princeton, N.J.

—— (1966) Paul J. Cohen and the continuum problem. *Proceedings of the International Congress of Mathematicians, Moscow, August 16-26, 1966*, Izdatél'stvo 'MIR', Moscow, 1968, 15-20.

Cohen, Paul J. (1963-4) The independence of the continuum hypothesis, and ibid., *II. Proceedings of the National Academy of Sciences*, 50 (1963), 1143-8 and 51 (1964) 105-110.

Davis, Martin (ed.), (1965) *The undecidable. Basic papers on undecidable propositions, unsolvable problems and computable functions*. Raven Press, Hewlitt, N.Y.

Dedekind, Richard (1888) *Was sind und was sollen die Zahlen?* Braunschweig, 6th edition, 1930.

Fraenkel, Abraham A. (1922) Der Begriff 'definit' und die Unabhängigkeit des Auswahlaxioms. *Sitzungsberichte der Preussische Akademie der Wissenschaften, Physikalische-mathematische Klasse*, 1922, 253-7.

Gentzen, Gerhard (1936) Die Widerspruchsfreiheit der reinen Zahlentheorie. *Mathematische Annalen*, 112, 493-565.

Gödel, Kurt (1930) Die Vollständigkeit der Axiome des logischen Funktionenkalküls. *Monatshefte für Mathematik und Physik*, 37, 349-60.

—— (1931) Über formal unentscheidbare Sätze der Principia Mathematica und verwandter Systeme, I. Ibid., 38, 173-98.

—— (1931-2) Über Vollständigkeit und Widerspruchsfreiheit, *Ergebnisse eines mathematischen Kolloquiums*, Heft 3 (for 1930-31, published 1932), 12-13.

—— (1932) Zum intuitionistischen Aussagenkalkül, *Akademie der Wissenschaften in Wien. Mathematisch-naturwissenschaftliche Klasse, Anzeiger*, 69, 65-6.

—— (1932-3) Zur intuitionistischen Arithmetik und Zahlentheorie. *Ergebnisse eines mathematischen Kolloquiums*, Heft 4 (for 1931-3, published 1933), 34-8.

—— (1934) *On undecidable propositions of formal mathematical systems*. Notes by S.C. Kleene and Barkley Rosser on lectures at the

Institute for Advanced Study, 1934, Princeton, N.J., 30 pp. Reprinted in Davis (1965).

—— (1934a) Review of Skolem (1933). *Zentralblatt für Mathematik und ihre Grenzgebiete*, 7, 193-4.

—— (1936) Über die Länge von Beweisen. *Ergebnisse eines mathematischen Kolloquiums*, Heft 7 (for 1934-5, published 1936), 23-4.

—— (1938) The consistency of the axiom of choice and of the generalized continuum-hypothesis. *Proceedings of the National Academy of Sciences*, 24, 556-7.

—— (1939) Consistency-proof for the generalized continuum-hypothesis. Ibid., 25, 220-4.

—— (1940) *The consistency of the axiom of choice and of the generalized continuum-hypothesis with the axioms of set theory.* Notes by George W. Brown on lectures at the Institute for Advanced Study, autumn term 1938-9. *Annals of Mathematics Studies*, no. 3, Princeton University Press, Princeton, N.J., 66 pp. Reviewed (at the invitation of J.D. Tamarkin) by S.C. Kleene in *Mathematical Reviews*, 2 (1941), 66-7.

—— (1947) What is Cantor's continuum problem? *American Mathematical Monthly*, 54, 515-25.

—— (1958) Über eine bisher noch nicht benützte Erweiterung des finiten Standpunktes. *Dialectica*, 12 (47/48), 280-7.

Henkin, Leon (1947) The completeness of formal systems. Ph.D. thesis, Princeton.

—— (1950) Completeness in the theory of types. *Journal of Symbolic Logic*, 15, 81-91.

Heyting, Arend (1934) *Mathematische Grundlagenforschung, Intuitionismus, Beweistheorie.* Ergebnisse der Mathematik und ihrer Grenzgebiete, 3 (4), Springer, Berlin.

Hilbert, David (1900) Mathematische Probleme, Vortrag, gehalten auf dem Internationalen Mathematiker-Kongress zu Paris 1900. *Nachrichten von der Königlichen Gesellschaft der Wissenschaften zu Göttingen*, 253-97. French translation with emendations and additions, *Compte rendu du Deuxième Congrès International des Mathématiciens tenu à Paris du 6 au 12 août 1900*, Gauthier-Villars, Paris, 1902, 58-114.

—— (1904) Über die Grundlagen der Logik und der Arithmetik. *Verhandlungen des Dritten Internationalen Mathematiker-Kongresses in Heidelberg vom 8. bis 13. August 1904.* Trubner, Leipzig, 1905, 174-85.

—— (1925) Über das Unendliche. Address at Münster, June 4, 1925 (Westphalian Mathematical Society), *Mathematische Annalen*, 95 (1926), 161-90.

—— (1927) Die Grundlagen der Mathematik. Address at Hamburg July 1927, *Abhandlungen aus dem Mathematischen Seminar der Hamburgischen Universität*, 6 (1928), 65-85.

Hilbert, David and Ackermann, Wilhelm (1928) *Grundzüge der theoretischen Logik.* Springer, Berlin.

Hilbert, David and Bernays, Paul (1939) *Grundlagen der Mathematik*, vol. 2, Springer, Berlin.

Kleene, Stephen C. (1935) A theory of positive integers in formal logic.

*American Journal of Mathematics*, 57, 153-73, 219-44.

—— (1936) General recursive functions of natural numbers. *Mathematische Annalen*, 112, 727-42.

—— (1936a) λ-definability and Recursiveness. *Duke Mathematical Journal*, 2, 340-53.

——— (1943) Recursive predicates and quantifiers. *Transactions of the American Mathematical Society*, 53, 41-73.

—— (1952) *Introduction to metamathematics*, North-Holland, Amsterdam, Noordhoff, Groningen, and Van Nostrand, Princeton, New York and Toronto, Seventh reprint, Wolters-Noordhoff, Groningen, North-Holland, Amsterdam-Oxford, American Elsevier, New York, 1974.

—— (1959) Recursive functionals and quantifiers of finite types I, *Transactions of the American Mathematical Society*, 91, 1-52.

—— (1967) *Mathematical logic*, John Wiley and Sons, New York, London, Sydney.

Löwenheim, Leopold (1915) Über Möglichkeiten im Relativkalkül. *Mathematische Annalen*, 76, 447-70.

Markov, Andrei Andreevič (1951) Theory of algorithms. *Translations of the American Mathematical Society* (2), 15 (1960), 1-14, Russian original, 1951.

Peano, Guisseppe (1889) *Arithmetices principia, nova methodo exposita*, Bocca, Turin.

Post, Emil Leon (1936) Finite combinatory processes — formulation I. *Journal of Symbolic Logic*, 1, 103-5.

—— (1943) Formal reductions of the general combinatorial decision problem. *American Journal of Mathematics*, 65, 197-215.

Rosser, Barkley (1936) Extensions of some theorems of Gödel and Church. *Journal of Symbolic Logic*, 1, 87-91.

Sheperdson, J.C. and Sturgis, H.E. (1963) The compatability of partial recursive functions. *Journal of the Association for Computing Machinery*, 10, 217-55.

Skolem, Thoralf (1920) Logisch-kombinatorische Untersuchungen über die Erfüllbarkeit oder Beweisbarkeit mathematischer Sätze nebst einem Theoreme über dichte Mengen. *Skrifter utgit av Videnskapsselskapet i Kristiania*, I, *Matematisk-naturvidenskabelig klasse* 1920, 4, 36 pp.

—— (1922) Einige Bemerkungen zur axiomatischen Begründung der Mengenlehre. *Wissenschaftliche Vorträge gehalten auf dem Fünften Kongress der Skandinavischen Mathematiker in Helsingfors vom 4. bis 7. Juli 1922*, Helsingfors, 1923, 217-32.

—— (1923) Begründung der elementaren Arithmetik durch die rekurrierende Denkweise ohne Anwendung scheinbare Veränderlichen mit unendlichem Ausdehnungsbereich. *Skrifter utgit av Videnskapsselskapet i Kristiania*, I. *Matematisk-naturvidenskabelig klasse 1923*, 6, 38 pp.

—— (1933) Über die Unmöglichkeit einer vollständigen Charakterisierung der Zahlenreihe mittels eines endlichen Axiomensystems. *Norsk matematisk forenings skrifter*, ser. 2, 10, 73-82.

—— (1934) Über die Nicht-charakterisierbarkeit der Zahlenreihe mittels endlich oder abzählbar unendlich vieler Aussagen mit ausschliesslich Zahlenvariablen. *Fundamenta Mathematicae*, 23, 150-61.

Tarski, Alfred (1933) Der Wahrheitsbegriff in den formalisierten Sprachen. *Studia Philosophica*, 1 (1936), 261-405 (offprints dated 1935). Translated from Polish original 1933.

—— (1956) *Logic, semantics, metamathematics, papers from 1923 to 1938*, Clarenden, Oxford.

Turing, Alan Mathison (1936-7) On computable numbers, with an application to the Entscheidungsproblem. *Proceedings of the London Mathematical Society*, ser. 2, 42, 230-65.

van Heijenoort, Jean (ed.) (1967) *From Frege to Gödel, a source book in mathematical logic, 1879-1931*, Harvard University Press, Cambridge.

von Neumann, John (1923) Zur Einführung der transfiniten Zahlen. *Acta litterarum ac scientiarum Regiar Universitatis Hungaricae Francisco-Josephinae, Sectio scientiarum mathematicarum*, 1, 199-208.

Weyl, Hermann (1910) Über die Definitionen der mathematischen Grundbegriffe. *Mathematisch-naturwissenschaftliche Blätter*, 7, 93-5, 109-13.

—— (1918) *Das Kontinuum. Kritische Untersuchungen über die Grundlagen der Analysis*. Gruyter, Leipzig.

Whitehead, Alfred North and Russell, Bertrand (1910-13) *Principia mathematica*, 1, 1910; 2, 1912; 3, 1913. University Press, Cambridge, England.

Zermelo, Ernst (1904) Beweiss, dass jede Menge wohlgeordnet werden kann. *Mathematische Annalen*, 59, 514-16.

—— (1908) Neuer Beweiss für die Möglichkeit einer Wohlordnung, Ibid., 65, 107-28.

—— (1908a) Untersuchungen über die Grundlagen der Mengenlehre I. Ibid., 261-81.

# AN ADDENDUM TO
# 'THE WORK OF KURT GÖDEL'[1]

Gödel has called to my attention that p. 64 is misleading in regard to the discovery of the finite axiomatization and its place in his proof of the consistency of GCH. For the version in (1940), as he says on p. 1, 'The system $\Sigma$ of axioms for set theory which we adopt [a finite one] ... is essentially due to P. Bernays ...'[2] However, it is not at all necessary to use a finite axiom system. Gödel considers the more suggestive proof to be the one in (1939), which uses infinitely many axioms.[3]

His main achievement regarding the consistency of GCH, he says, really is that he first introduced the concept of constructible sets into set theory defining it as in (1939), proved that the axioms of set theory (including the axiom of choice) hold for it, and conjectured that the continuum hypothesis also will hold. He told these things to von Neumann during his stay at Princeton in 1935. The discovery of the proof of this conjecture on the basis of his definition is not too difficult. Gödel gave the proof (also for GCH) not until three years later because he had fallen ill in the meantime. This proof was using a submodel of the constructible sets in the lowest case countable, similar to the one commonly given today.

Regarding p. 50: When Gödel wrote his completeness paper (1930), he did not know of Skolem's work (1922), which was not mentioned in Hilbert–Ackermann (1928).

Corresponding to pp. 51-6: The proof in Gödel's undecidability paper (1931) may be thought rather awkward. But the explanation lies in the fact that it is completely formalized.[4]

## NOTES

1. Stephen C. Kleene, *The work of Kurt Gödel*, pp. 48-71 above.
2. Paul Bernays, A system of axiomatic set theory — Part I. *Journal of Symbolic Logic*, 2 (1937), 65-77.
3. The error on p. 64 lines 20-1 is not original with me. In 1975, instead of rereading (1940) from which I learned the subject, or better reading (1939), I used material I had just taught in class from a text by a set-theorist which states, incorrectly, 'Gödel was the first to realize that the Comprehension Schema can be replaced by a finite number of its instances ...'.

4. Gödel has discussed extensively the conceptual framework of his work in letters and personal communications published in pp. 9-12, 84-86, 186, 189-190, 324-326 of Hao Wang, *From mathematics to philosophy*, Routledge & Kegan Paul, London, Humanities Press, New York, 1974, xiv + 428 pp. Also comments by Gödel are in Davis (1965), van Heijenoort (1967) and Abraham Robinson, *Non-standard analysis*, 2nd edn, North-Holland, Amsterdam, 1974, xii + 293 pp.

# IV

# The Reception of Gödel's Incompleteness Theorems

John W. Dawson, Jr.

Die Arbeit über formal unentscheidbare Sätze wurde wie ein Erdbeben empfunden; insbesondere auch von Carnap (Popper, 1980).

Kurt Gödel's achievement in modern logic ... is a landmark which will remain visible far in space and time (John von Neumann).[1]

It is natural to invoke geological metaphors to describe the impact and the lasting significance of Gödel's incompleteness theorems. Indeed, how better to convey the impact of those results — whose effect on Hilbert's Programme was so devastating and whose philosophical reverberations have yet to subside — than to speak of tremors and shock waves? The image of shaken foundations is irresistible.

Yet to adopt such seismic imagery is to suggest that the aftermath of intellectual upheaval is comparable to that of geological cataclysm: a period of slow rebuilding, preceded initially by widespread confusion and despair, or perhaps determined resistance; and though we might expect Gödel's discoveries to have provoked just such reactions, according to most commentators they did not. Thus van Heijenoort states 'although [Gödel's paper] caused some momentary surprise, its results were soon widely accepted' (van Heijenoort, 1967, p. 594). Similarly, Kreisel has averred that 'expected objections never materialized' (Kreisel, 1979, p. 13), and Kleene, speaking of the second incompleteness theorem (whose proof was only sketched in Gödel's (1931)), has even claimed 'it seems no one doubted it' (Kleene 1976, p. 767).

If these accounts are correct, one of the most profound discoveries in the history of logic and mathematics was assimilated promptly and almost without objection by Gödel's contemporaries — a circumstance so remarkable that it demands to be accounted for. The received explanation seems to be that Gödel, sensitive to the philosophical climate of opinion and anticipating objections to his work, presented his results with such clarity and rigour as to render them incontestable, even at a time of fervid debate among competing mathematical philosophies. The sheer force of Gödel's logic, as it were, swept away opposition so effectively that Gödel abandoned his stated intention of publishing a detailed proof of the second theorem (1931, p. 198).

On the last point there can be no dispute, as Gödel stated explicitly to van Heijenoort that 'the prompt acceptance of his results was one of the reasons that made him change his plan' (van Heijenoort, 1967, footnote 68a, p. 616). We may question, however, to what extent Gödel's subjective impression reflected objective circumstances. We must also recognize the hazard in assessing the cogency of Gödel's arguments from our own perspective. To be sure, the exposition in (Gödel, 1931) *now* seems clear and compelling; the proofs strike us as detailed but not intricate. But did it seem so at the time? After all, arithmetization of syntax was then a novel device, and logicians were not then so accustomed to the necessity for making precise distinctions between object- and metalanguage. Indeed, J. Barkley Rosser (who himself contributed to the improvement of Gödel's results) has observed that only *after* Gödel's paper appeared did logicians realize how careful they had to be (cf. Grattan-Guinness, 1981, footnote 3, p. 499); precisely *because* of Gödel's results we no longer share the formalists' naive optimism, and so we are likely to be more receptive to Gödel's ideas.

Our faith in the efficacy of logic as a tool for overcoming intellectual resistance should also be tempered by consideration of the reception accorded other 'paradoxical' results. For example, the Löwenheim–Skolem theorem, first enunciated by Löwenheim in 1915 and established with greater precision and broader generality by Skolem in a series of papers from 1920 to 1929, is certainly less profound than Gödel's discovery (with which, however, it is still sometimes confused) — yet it generated widespread misunderstanding and bewilderment, even as late as December 1938. (See, in particular, the discussion in Gonseth, 1941, pp. 47-52.)

In what follows, I shall examine the reaction to Gödel's theorems in some detail, with the aim of showing that there *were* doubters and critics, as well as defenders and rival claimants to priority. (Of course, there were also some who *accepted* Gödel's results without fully *understanding* them.)

## 1. 1930: ANNOUNCEMENT AT KÖNIGSBERG

Elsewhere (Dawson, 1984) I have described in detail the circumstances surrounding Gödel's first public announcement of his incompleteness discovery. In summary, the event occurred during a discussion on the foundations of mathematics that took place in Königsberg, 7 September 1930, as one of the final sessions of the Second Conference on Epistemology of the Exact Sciences organized by the *Gesellschaft für empirische Philosophie*. At the time, Gödel was virtually unknown outside Vienna; he had come to the conference to deliver a twenty-minute talk on the results of his dissertation, completed the year before and just then about to appear in print. In that work, Gödel had established a result of prime importance for the advancement of Hilbert's programme: the completeness of first-order logic (or, as it was then called, the restricted functional calculus); so it could hardly have been expected that the day after his talk Gödel would suddenly undermine that programme by asserting the existence of formally undecidable propositions. As Quine has remarked, '[although] completeness was expected, an actual proof of completeness was less expected, and a notable accomplishment. It came as a welcome reassurance. ... On the other hand the incompletability of elementary number theory came as an upset of firm preconceptions' (Quine, 1978, p. 81).

To judge from the (edited) transcript of the discussion published in *Erkenntnis* (Hahn et al., 1931), none of the other participants at Königsberg had the slightest inkling of what Gödel was about to say, and the announcement itself was so casual that one suspects that some of them may not have realized just what he *did* say. In particular, the transcript gives no indication of any discussion of Gödel's remarks, and there is no mention of Gödel at all in Reichenbach's post-conference survey of the meeting (published in *Die Naturwissenschaften* 18: 1093-4). Yet two, at least, among those present *should* have had foreknowledge of Gödel's results: Hans Hahn and Rudolf Carnap.

Hahn had been Gödel's dissertation adviser. He chaired the discussion at Königsberg, and it was presumably he who invited Gödel to take part. Of course, Gödel may not have confided his new discovery to him — indeed, Wang has stated that 'Gödel completed his dissertation before showing it to Hahn' (Wang, 1981, footnote 2, p. 654) — but in introductory remarks to the dissertation that were *deleted* from the published version (at whose behest we do not know), Gödel had explicitly raised the possibility of incompleteness (without claiming to have demonstrated it). Perhaps Hahn just didn't take the possibility seriously.

In any case, Gödel *did* confide his discovery to Carnap prior to the discussion at Königsberg, as we know from *Aufzeichnungen* in Carnap's *Nachlass*. Specifically, on 26 August 1930, Gödel met Carnap, Feigl, and Waismann at the Cafe Reichsrat in Vienna, where they discussed their travel plans to Königsberg. Afterward, according to Carnap's entry for that date, the discussion turned to 'Gödels Entdeckung: Unvollständigkeit des Systems der PM; Schwierigkeit des Widerspruchsfreiheitbeweises'.[2] Three days later another meeting took place at the same cafe. On that occasion, Carnap noted 'Zuerst [before the arrival of Feigl and Waismann] erzählt mir *Gödel* von seiner Entdeckungen.' Why then at Königsberg did Carnap persist in advocating consistency as a criterion of adequacy for formal theories? That he might have done so just to provide an opening for Gödel seems hardly credible. It seems much more likely that he simply failed to *understand* Gödel's ideas. (As it happens, a subsequent note by Carnap dated 7 February 1931, after the appearance of Gödel's paper, provides confirmation: 'Gödel hier. Über seine Arbeit, ich sage, dass sie doch schwer verständlich ist.') Later, of course, Carnap was among those who helped publicise Gödel's work; but Popper's remark quoted at the head of this article seems to be an accurate characterisation of Carnap's initial reaction.

One of the discussion participants who did immediately appreciate the significance of Gödel's remarks was John von Neumann: after the session he drew Gödel aside and pressed him for further details. Soon thereafter he returned to Berlin, and on 20 November he wrote Gödel to announce his discovery of a remarkable [*bemerkenswert*] corollary to Gödel's results: the unprovability of consistency. In the meantime, however, Gödel had himself discovered his second theorem and had incorporated it into the text of his paper; the finished article was received by the editors of *Monatshefte* November 17.[3]

## 2. 1931: PUBLICATION AND CONFRONTATION

In January 1931 Gödel's paper was published. But even before then, word of its contents had begun to spread. So, for example, on 24 December 1930, Paul Bernays wrote Gödel to request a copy of the galleys of (1931), which Courant and Schur had told him contained 'bedeutsamen und überraschenden Ergebnissen'. Gödel responded immediately, and on 18 January 1931, Bernays acknowledged his receipt of the galleys in a 16-page letter in which he described Gödel's results as 'ein erheblicher Schritt vorwärts in der Erforschung der Grundlagenprobleme'.

The Gödel–Bernays correspondence is of special interest (in the absence of more direct evidence[4]) for the light it sheds on Hilbert's reaction. In the same letter of 18 January, Bernays discussed Hilbert's 'recent extension of the usual domain of number theory' — his introduction of the $\omega$-rule — and von Neumann's belief that every finitary method of proof could be formalized in Gödel's system P. Bernays himself saw Gödel's theorem as establishing a disjunction: either von Neumann was right, and no finitary consistency proof was possible for the systems Gödel considered, or else some finitary means of proof were not formalizable in P — a possibility Gödel had expressly noted in his paper. In any case, Bernays felt, one was impelled [*hingedrängt*] to weaken Gödel's assumption that the class of axioms and the rules of inference be (primitively) recursively definable. He suggested that the system obtained by adjoining the $\omega$-rule would escape incompleteness yet might be proved consistent by finitary means. But (according to Carnap's *Aufzeichnung* of 21 May 1931) Gödel felt that Hilbert's programme would be compromised by acceptance of the $\omega$-rule.

Somewhat later Gödel sent Bernays an offprint of the incompleteness paper, enclosing a copy for Hilbert as well. In his acknowledgement of 20 April, Bernays confessed his inability to see why a truth predicate could not be formally defined in number theory — he went so far as to propose a candidate for such a definition — and why Ackermann's consistency proof (which he had accepted as correct) could not be formalized there as well. By 3 May, when he wrote Gödel once more, Bernays had recognised his errors, but the correspondence remains of interest, not only because it exposes Bernays' difficulties in assimilating the consequences of Gödel's theorems, but because it furnishes independent evidence of Gödel's awareness of the formal undefinability of the notion of truth — a fact nowhere mentioned in (1931).[5]

Gödel's formal methods, as employed in the body of (1931), thus seem to have served their purpose in securing the acceptance of his results by three of the leading formalists. But at the same time, even among those who appreciated the value of formalization, Gödel's precise specification of the system P raised doubts as to the *generality* of his conclusions.[6] On the other hand, those opposed to formal systems could point to Gödel's results as reason for dismissing such systems altogether.

Partly to obviate such objections, Gödel soon extended his results to a wider class of systems (in (1930/31) and (1934)), and in the introduction to (1931) he also gave *informal* proofs of his results based on the soundness of the underlying axiom systems rather than on their formal consistency properties. Undoubtedly, he hoped this informal précis would help his readers to cope with the formal detail to follow; but all too many read no further. Ironically, because of the misinterpretations it subsequently spawned, the introduction to (1931) has been called that paper's 'one "mistake"' (Helmer, 1937).

During 1931, Gödel spoke on his incompleteness results on at least three occasions: at a meeting of the Schlick circle (15 January), in Karl Menger's mathematics colloquium (22 January), and, most importantly, at the annual meeting of the *Deutsche Mathematiker–Vereinigung* in Bad Elster (15 September), where, in Ernst Zermelo, he encountered one of his harshest critics.

At issue between the two men were profound differences in philosophy and methodology. In his own talk at Bad Elster (abstracted in Zermelo, 1932), Zermelo lashed out against 'Skolemism, the doctrine that *every* mathematical theory, even set theory, is realizable in a countable model' — a doctrine that Zermelo regarded as an embodiment of Richard's antinomy.

For Zermelo, quantifiers were infinitary conjunctions or disjunctions of unrestricted cardinality; and since the truth values of compound statements were therefore determined by transfinite induction on the basis of the truth values assigned to the *Grundrelationen*, Zermelo argued that *this determination itself* constituted the proof or refutation of each proposition. There were no 'undecidable' propositions, simply because Zermelo's infinitary logic had no syntactic component. Consequently, Zermelo dismissed Gödel's 'attempt' to exhibit undecidable propositions, saying that Gödel 'applied the "finitistic" restriction only to the "provable" statements' of his system, 'not to *all* statements belonging to it',[7] so that Gödel's result, like the Löwenheim–Skolem theorem, depended on (unwarranted) cardinality restrictions. It said nothing about the existence of 'absolutely unsolvable problems in mathematics'.

In his published remarks, Zermelo did not fault the correctness of Gödel's argument; he merely took it as evidence of the untenability of the "finitistic" restriction'. But on 21 September, immediately following the conference, Zermelo wrote privately to Gödel to inform him of 'an essential gap' [*eine wesentliche Lücke*] in his argument. (See Dawson (1985) for the full text of this letter.) Indeed, Zermelo argued, simply by omitting the proof predicate from Gödel's construction one would obtain a formal sentence asserting its own falsity, yielding thereby 'a *contradiction* analogous to Russel[l]'s antinomy'. Zermelo's letter prompted a further exchange of letters between the two men (published in Grattan-Guinness (1979)), with Gödel patiently explaining the workings of his proof, pointing out the impossibility of defining truth combinatorially within his system, and emphasising that the introductory pages of his paper (to which Zermelo had referred) did not pretend to precision, as did the detailed considerations later on. In reply, Zermelo thanked Gödel for his letter, from which, he said, he had gained a better understanding of what Gödel meant to say; but in his published report he still failed utterly to appreciate Gödel's distinctions between syntax and semantics.

## 3. RECOGNITION, AND CHALLENGES TO PRIORITY

Despite (or perhaps because of) Zermelo's reputation as a polemicist, his crusade against 'Skolemism' won few adherents, and his criticisms of Gödel's work seem to have been disregarded. In the spring of 1932 Karl Menger became the first to expound the incompleteness theorem to a popular audience, in his lecture *Die neue Logik* (published by Franz Deuticke in 1933 as one of 'fünf Wiener Vorträge' in the booklet *Krise und Neuaufbau in den exakten Wissenschaften*; see Menger (1978) for an English translation). The following June Gödel submitted (1931) to the University of Vienna as his *Habilitationsschrift*, and in January 1933 he accepted an invitation to spend the academic year 1933-34 at the newly established Institute for Advanced Study in Princeton.

On 11 March 1933, Gödel's *Dozentur* was granted. That same day Paul Finsler in Zürich addressed a letter to Gödel requesting a copy of (1931). He was interested in Gödel's work, he said, because it seemed to be closely related to earlier work of his own (Finsler, 1926). He had already glanced fleetingly at Gödel's paper, and he acknowledged that Gödel had employed 'a narrower and therefore sharper formalism'. It was 'of course of value', he conceded, 'actually to carry out [such] ideas in a specific formalism', but he had refrained from doing so because he felt that he had already established the result in a way that 'went further in its application to Hilbert's programme'.

Gödel recognised the challenge for what it was, and in his reply of March 25 he characterised Finsler's system as 'not really defined at all' [*überhaupt nicht definiert*], declaring that Finsler's ideas could not be carried out in a genuinely formal system, since the antidiagonal sequence defined by Finsler (on which his undecidable proposition was based) would never be representable within that same system. Finsler retorted angrily that it was not necessary for a system to be 'sharply' defined in order to make statements about it; it was enough that 'one be able to accept it as given and to recognize a few of its properties'. He could, he said, 'with greater justice' object to Gödel's proof on the grounds that Gödel had not shown the Peano axioms that he employed to be consistent (Gödel's second theorem notwithstanding!). He saw that the truth of Gödel's undecidable sentence could only be established meta-mathematically and so concluded that there was 'no difference in principle' between his and Gödel's examples.

Van Heijenoort has analysed Finsler's paper in detail, concluding that it 'remains a sketch' whose 'affinity [with Gödel's paper] should not be exaggerated' (1967, pp. 438-40). In effect, whereas 'Gödel put the notion of formal system at the very center of his investigations,' Finsler rejected such systems as artificially restrictive (prompting Alonzo Church to remark dryly that such 'restricted' notions have 'at least the merit of being precisely communicable from one person to another') (Church, 1946). Instead, especially in his (1944), Finsler attempted to show the consistency of the assertion that there were no 'absolutely undecidable' propositions.

In contrast, Emil Post directed his efforts toward showing that there *were* absolutely unsolvable problems in mathematics. Early on, nearly a decade before Gödel, Post realised that his methods could be applied to yield a statement undecidable within *Principia* whose truth could nevertheless be established by metamathematical considerations. Hence, Post concluded, 'mathematical proof was [an] essentially creative [activity]' whose proper elucidation would require an analysis of 'all finite processes of the human mind'; and since the implications of such an analysis could be expected to extend far beyond the incompleteness of *Principia*, Post saw no reason to pursue the latter.[8]

Concern for the question of *absolute* undecidability thus led Finsler, Post and Zermelo, in varying directions, away from consideration of particular formal systems. Unlike Finsler and Zermelo, however, Post expressed 'the greatest admiration' for Gödel's work, and he never sought to diminish Gödel's achievement. Indeed, Post acknowledged to Gödel that nothing he had done 'could have replaced the splendid actuality of your proof', and that 'after all it is not ideas but the execution of ideas that constitute[s] ... greatness.' Post's 'Account of an anticipation' was not submitted until 1941 and (after being rejected) only finally appeared in print in 1965, eleven years after Post's death.[9]

It is worth noting Gödel's own opinion of the notion of absolute undecidability, as expressed in the unsent letter draft cited above in footnote 5:

As for work done earlier about the question of formal decidability of mathematical propositions, I know only a paper by Finsler .... However, Finsler omits exactly the main point which makes a proof possible, namely restriction to some well-defined formal system in which the proposition is undecidable. For he had the nonsensical aim of proving formal undecidability in an absolute sense. This leads to [his] nonsensical definition [of a system of signs and of formal proofs therein] ... and to the flagrant inconsistency that he decides the 'formally undecidable' proposition by an argument ... which, *according to his own definition ... is a formal proof.* If Finsler had confined himself to some well-defined formal system S, his proof ... could [with the hindsight of Gödel's own methods] be made correct and applicable to any formal system. I myself did not know his paper when I wrote mine, and other mathematicians or logicians probably disregarded it because it contains the obvious nonsense just mentioned (Gödel to Yossef Balas, 27 May 1970).

Especially as applied to Finsler's later work, the epithet 'obvious nonsense' is well-deserved — (Finsler 1944) in particular is an almost pathological example of the confusion that can arise from failure to distinguish between use and mention — and Gödel was undoubtedly sensitive to any rival claim to priority for his greatest discovery. Nevertheless, the passage quoted above is uncharacteristically harsh, and insofar as it ridicules the idea of 'proving formal undecidability in an absolute sense' it seems to ignore Gödel's own remarks in 1946 before the Princeton Bicentennial Conference, in which he suggested that, despite the incompleteness theorems, 'closer examination shows that [such negative] results do not make a definition of the absolute notions concerned impossible under all circumstances.' In particular, he noted that 'by a kind of miracle' there is an absolute definition of the concept of computability, even though 'it is merely a special kind of demonstrability or decidability'; and in fact, the incompleteness theorems may be subsumed as corollaries to the existence of algorithmically unsolvable problems. Finsler, of course, had no such thing in mind in 1933, nor did Gödel. But what of Post? Had he not been hampered by manic-depression, might he have pre-empted Gödel's results? In any case, Gödel contributed far less than Post to the development of recursion theory, even though he gave the first definition of the notion of general recursive function (in (1934), following a suggestion of Herbrand). Until Turing's work (1936/37), Gödel resisted accepting Church's thesis (see Davis (1982) for a detailed account), and we may well wonder whether Gödel's focus on specific formalisms did not tend to blind *him* to the larger question of algorithmic undecidability.

Gödel repeatedly stressed the importance of his philosophical outlook to the success of his mathematical endeavors; perhaps it may also have been responsible for an occasional oversight. (See Kleene (1981a) and (1981b) for recent accounts of the origins of recursive function theory and Feferman (1985) for another view of Gödel's role as a bystander.)

## 4. ASSIMILATION AND LATER CRITICISM

After his IAS lectures in the winter and spring of 1934, Gödel returned to Vienna and turned his attention to set theory. By the fall of 1935, when he returned briefly to the IAS, he had already succeeded in proving the relative consistency of the axiom of choice. In the meantime, however, he had also once had to enter a sanatorium for treatment of depression. Back in America, he suffered a relapse that forced him to return to Austria just two months after his arrival. He re-entered a sanatorium and did not resume teaching at the University of Vienna until the spring of 1937.

During this period of incapacitation, Gödel's incompleteness results were improved by Rosser (1936) (who weakened the hypothesis of $\omega$-consistency to that of simple consistency) and extended, as already noted, through the development of recursion theory by Church, Kleene and Turing. By 1936, then, the incompleteness theorems would seem to have taken their place within the corpus of firmly established mathematical facts.

In that year, however, the correctness of Gödel's conclusions was challenged in print by Charles Perelman (1936), who asserted that Gödel had in fact discovered an *antinomy*. According to Perelman, if the soundness of the underlying axiom system were assumed (as Gödel had done in his informal introductory remarks), Gödel's methods could be employed to prove two false equivalences, namely

$$\text{Dem} \, ( \sim \, _qFq) \equiv \text{Dem} \, (_qFq)$$

and

$$\text{Dem} \, (_qFq) \equiv \, \sim \text{Dem} \, (_qFq),$$

where '$Fq$' denotes Gödel's undecidable sentence and 'Dem'

denotes Gödel's provability predicate (called 'Bew' in Gödel (1931)). The second statement is obviously a contradiction, and Perelman proposed to exorcise it by the radical expedient of rejecting the admissibility of the set of Gödel numbers of unprovable sentences.

Perelman's equivalences display a rather obvious conflation of object- and metalanguage, and, as Kleene noted in his review of (Perelman, 1936), if expressed without abuse of notation, 'the first of them is not false, and the second is not deducible' (Kleene, 1937a) — indeed, properly formulated, the first equivalence is just the statement that '$Fq$' is formally undecidable. It would thus be tempting to dismiss Perelman as a crank, were it not that his arguments were apparently taken quite seriously by many within the mathematical community — so much so, in fact, that two individuals, Kurt Grelling and Olaf Helmer, felt obliged to come to Gödel's defence.

Grelling was among those who had travelled with Gödel to Königsberg in 1930. On 2 February 1937, he wrote Gödel to advise him of Hempel's report that 'angesehene Mathematiker' in Brussels and Paris were being 'taken in' [*hereingefallen*] by Perelman's arguments. If Gödel were not already planning to publish a rebuttal, Grelling requested permission to do so on his behalf. Gödel did not in fact enter into the controversy, and Grelling's article (1937/38) appeared at about the same time as Helmer's (1937).

Grelling began his reply to Perelman by giving an informal outline of Gödel's proof, drawing a careful distinction between arithmetical statements and their metamathematical counterparts. Through the process of arithmetisation, he explained, a metamathematical statement Q' was associated to Gödel's arithmetical statement Q, in such a way that to any formal proof of Q there would correspond a metamathematical proof of Q'; but, he asserted, a *metamathematical* proof or refutation of Q' would lead directly to a *metamathematical* contradiction ('Sowohl ein Beweis von Q' als auch eine Widerlegung würde unmittelbar auf einen Widerspruch führen') (Grelling, 1937/38, p. 301), and so Q itself must be formally undecidable. Grelling went on to analyse Perelman's formal arguments, especially his claims that

$$(n) . \mathrm{Dem} \,(_nFn) . \supset . \mathrm{Dem} \,(\mathrm{Dem} \,(_nFn)) \qquad (1)$$

and

$$(n) . \sim \text{Dem} \left( {}_nFn \right) . \supset . \text{Dem} \left( \sim \text{Dem} \left( {}_nFn \right) \right) \qquad (2)$$

were provable in Gödel's system. Without commenting on the abuse of notation involved (whereby 'Dem' should correspond to Gödel's *meta*mathematical predicate 'Bew', not to the Gödel number of its arithmetical counterpart), he (correctly) accepted (1) as demonstrable, while rejecting (2).

Helmer, on the other hand, stressed Perelman's notational confusion, pointing out that '"Dem" is a predicate applicable to numbers only,' so that 'if formula (1) is to be significant, the expression "$Fn$" must denote a number and not a sentence'. One might therefore construe '$Fn$' 'as a designation of the number correlated [via arithmetisation] to the sentence resulting from the substitution of "$n$" at the argument-place of the $n$th predicate in the syntactical denumeration of all the numerical predicates', but even so, Helmer argued, formula (1) '[could] not possibly be legitimated' (Helmer, 1937).

The articles of Grelling and Helmer were criticised in their turn by Rosser and Kleene in reviews published in the *Journal of Symbolic Logic* (Rosser (1938) and Kleene (1937b)). Rosser noted that '[Grelling's] exposition of the closing steps of Gödel's proof' did not agree with his own understanding of it, and that Grelling's statement quoted above to the effect that Q' was metamathematically undecidable was 'indubitably false, ... since in a preceding sentence he says that Q' is a metamathematical theorem'. Here, however, we may quibble: Grelling does *not* say that Q' is a metamathematical *theorem* — he refers to it merely as 'ein metamathematischer Satz', which in this context simply means a statement or proposition; rather, Grelling's statement is 'indubitably false' precisely because Q' *is* a metamathematical theorem (in the sense that metamathematical arguments show Q to be *true* in its intended interpretation). As to Helmer, Kleene agreed that the source of Perelman's errors lay in his failure to distinguish between formulas and their Gödel numbers, but he also noted Helmer's failure 'to distinguish as consistently between ... *metamathematical statements* and *formal mathematical sentences* as between those and ... *syntactical numbers*' of the latter. At the same time he affirmed the basic legitimacy of Perelman's equivalance (1), 'Helmer notwithstanding'.

The controversy sparked by Perelman's paper exposes not only the fragility of the earlier acceptance of Gödel's results, but also the misunderstandings of those results by their would-be defenders. Thus Grelling and Helmer, while criticising Perelman's confusion between object- and metalanguage, were themselves not always careful about syntactic distinctions. Grelling saw a metamathematical contradiction where none existed, and Helmer wrongly traced Perelman's error to a formula that is in fact provable (though not so easily as Perelman thought); while Rosser, by mistranslating 'Satz', failed to recognise the real source of Grelling's misunderstanding. Only Kleene emerged from the debate untarnished ('clean').

For a more detailed account of Perelman's claims and their refutation, the reader may turn to chapter 3, section 5 of Ladrière's book (1957). That source also includes a discussion of two other, still later objections to Gödel's work by Marcel Barzin (1940) and Jerzy Kuczyński (1938). Neither of their challenges seems to have received much notice at the time, so I shall devote little attention to them here, except to note that their criticisms, unlike Perelman's, were based on Gödel's detailed formal proof rather than his informal introductory arguments; like Perelman, however, both Barzin and Kuczyński thought Gödel had discovered an antinomy. (In essence, Barzin confused formal expressions with their Gödel numbers, while Kuczyński, to judge from Mostowski's review (1938), overlooked the formal antecedent 'Wid($\kappa$)' in Gödel's second theorem.)

## 5. LINGERING DOUBTS

In 1939, the second volume of Hilbert and Bernays' *Die Grundlagen der Mathematik* appeared, in which, for the first time, a complete proof of Gödel's second incompleteness theorem was given. Whether because of its meticulous treatment of syntactic details or because of Hilbert's implied imprimatur, the book at last seems to have stilled serious opposition to Gödel's work, at least within the community of logicians.[10]

Outside that community it is difficult to assess to what extent Gödel's results were known, much less accepted or understood. It seems likely that many working mathematicians either remained only vaguely aware of them or else regarded them as having little or no relevance to their own endeavors. Indeed, until the 1970s, with the work of Matijasevic and, later, of Paris and Harrington, number theorists could (and often did) continue to regard undecidable arithmetical statements as artificial contrivances of interest only to those concerned with foundations.

Among the few non-logicians (at least to my knowledge) who took cognizance of Gödel's work in writings of the period was Garrett Birkhoff. In the first edition of his *Lattice Theory* (1940) he observed (p. 128) that 'the existence of "undecidable" propositions ... seems to have been established by Skolem [n.b.] and Gödel'. In a footnote, however, he qualified even that tentative acceptance, noting that 'Such a conclusion depends of course on prescribing all admissible methods of proof ... [and] hence ... should be viewed with deep skepticism.' In particular, he believed that 'Carnap [had] stated plausible methods of proof excluded by Gödel'. In the revised edition (1948), the corresponding footnote (p. 194) is weakened to read 'The question remains whether there do not exist perfectly "valid" methods of proof excluded by [Gödel's] particular logical system'; but the reference to Skolem is retained in the text proper, and an added sentence (p. 195) declares that the proof of the existence of undecidable propositions 'is however non-constructive, and depends on admitting the existence of uncountably many "propositions", but only countably many "proofs"'. Clearly, Birkhoff had not actually read Gödel's paper. His statements echo those of Zermelo, but he does not cite Zermelo's report.

Of still greater interest are the reactions of two philosophers of stature: Ludwig Wittgenstein and Bertrand Russell. Wittgenstein's well-known comments on Gödel's theorem appear in Appendix I of his posthumously published *Remarks on the Foundations of Mathematics* (extracted in English translation on pp. 431-435 of Benacerraf and Putnam (1964)). Dated in the preface of the volume to the year 1938, they were never intended to be published and perhaps should not have been — but that, of course, is irrelevant to the present inquiry. Several commentators have discussed Wittgenstein's remarks in detail (see, for example, the articles by A.R. Anderson, Michael Dummett, and Paul Bernays, pp. 481-528 of Benacerraf and Putnam (1964)), and nearly all have considered them an embarrassment to the work of a great philosopher. Certainly it is hard to take seriously such objections as 'Why should not propositions ... of physics ... be written in Russell's symbolism?'; or 'The contradiction that arises when someone says "I am lying" ... is of interest only because it has tormented people'; or 'The proposition "P is unprovable" has a different sense afterwards [than] before it was proved' (Benacerraf and Putnam, 1964, pp. 432 and 434). Whether some more profound philosophical insights underlie such seemingly flippant remarks must be left for Wittgenstein scholars to debate; suffice it to say that in Gödel's opinion, Wittgenstein 'advance[d] a completely trivial and uninteresting misinterpretation' of his results (Gödel to Abraham Robinson, 2 July 1973).

As to Russell, two passages are of particular interest here, one published and one not, dating respectively from 1959 and 1963. The former, from *My Philosophical Development* (p. 114), forms part of Russell's own commentary on Wittgenstein's work:

> In my introduction to the *Tractatus*, I suggested that, although in any given language there are things which that language *cannot express* [my emphasis], it is yet always possible to construct a language of higher order in which these things can be said. There will, in the new language, still be things which it *cannot say* [my emphasis again], but which can be said in the next language, and so on *ad infinitum*. This suggestion, which was then new, has now become an accepted commonplace of logic. It disposes of Wittgenstein's mysticism and, I think, also of the newer puzzles presented by Gödel.

It might be maintained that in this passage Russell is making the same point that Gödel himself stressed in his (1930/31), that by passing to successively higher types one can obtain a transfinite sequence of formal systems such that the undecidable propositions constructed within each system are *decidable* in all subsequent systems. Contrary to Russell, however, Gödel emphasised that each of the undecidable propositions so constructed is already *expressible* at the lowest level.

The second passage is taken from Russell's letter to Leon Henkin of 1 April 1963:[11]

It is fifty years since I worked seriously at mathematical logic and almost the only work that I have read since that date is Gödel's. I realized, of course, that Gödel's work is of fundamental importance, but I was puzzled by it. It made me glad that I was no longer working at mathematical logic. If a given set of axioms leads to a contradiction, it is clear that at least one of the axioms must be false. Does this apply to school-boys' arithmetic, and, if so, can we believe anything that we were taught in youth? Are we to think that 2 + 2 is not 4, but 4.001? Obviously, this is not what is intended.

You note that we were indifferent to attempts to prove that our axioms could not lead to contradictions. In this, Gödel showed that we had been mistaken. But I thought that it must be impossible to prove that any given set of axioms does *not* lead to a contradiction, and, for that reason, I had paid little attention to Hilbert's work. Moreover, with the exception of the axiom of reducibility which I always regarded as a makeshift, our other axioms all seemed to me luminously self-evident. I did not see how anybody could deny, for instance, that q implies p or q, or that p or q implies q or p.

... In the later portions of the book ... are large parts consisting of ... ordinary mathematics. This applies especially to relation-arithmetic. If there is any mistake in this, apart from trivial errors, it must also be a mistake in conventional ordinal arithmetic, which seems hardly credible.

If you can spare the time, I should like to know, roughly, how, in your opinion, ordinary mathematics — or, indeed, any deductive system — is affected by Gödel's work.

A curious ambiguity infects this letter. Is Russell recalling his bewilderment at the time he first became acquainted with Gödel's theorems, or is he expressing his continuing puzzlement? Is he saying that, intuitively, he had recognised the futility of Hilbert's scheme for proving the consistency of arithmetic but had failed to consider the possibility of rigorously *proving* that futility? Or is he revealing a belief that Gödel had in fact shown arithmetic to be *in*consistent? Henkin, at least, assumed the latter; in response to Russell's closing request, he attempted to explain the import of Gödel's second theorem, stressing the distinction between incompleteness and inconsistency. Eventually a copy of Russell's letter made its way to Gödel, who remarked that 'Russell evidently misinterprets my result; however he does so in a very interesting manner ...' (Gödel to Abraham Robinson, 2 July 1973).

## 6. CONCLUSIONS

In so far as it refers to the acceptance of Gödel's results by *formalists*, the received view appears to be correct. Gödel's proofs dashed formalist hopes, but at the same time they were most persuasive to those committed to formalist ideals. In other quarters, the incompleteness theorems were by no means so readily accepted; objections were raised on both technical and philosophical grounds. Especially prevalent were the views that Gödel's results were antinomial or were of limited generality. Cardinality restrictions, in particular, were often perceived to be responsible for the phenomenon of undecidability.

Gödel succeeded where others failed because of his attention to syntactic and semantic distinctions, his restriction to particular formal systems, and his concern for relative rather than absolute undecidability. He anticipated resistance to his conclusions and took pains to minimise objections by his style of exposition and by his avoidance of the notion of objective mathematical truth (which was nevertheless central to his own mathematical philosophy). Though aware of criticisms of his work, he shunned public controversy and considered his results to have been readily accepted *by those whose opinion mattered to him*; nevertheless, his later extensions of his results display his concern for establishing their generality.

In the long run, the incompleteness theorems have led neither to the rejection of formal systems nor to despair over their limitations, but, as Post foresaw, to a reaffirmation of the creative power of human reason. With Gödel's theorems as centrepiece, the debate over mind versus mechanism continues unabated.

## NOTES

1. Remarks 14 March 1951, at the Einstein Award ceremonies.
2. 'Schwierigkeit', not 'Unmöglichkeit': we know from other

sources that Gödel did not obtain his second theorem until after the Königsberg meeting.

3. Even earlier, on 22 October, Gödel had communicated an abstract of his results (1930) to the Vienna Academy of Sciences.

4. In a letter to Constance Reid of 22 March 1966, Gödel stated that he 'never met Hilbert ... nor [had] any correspondence with him'. The stratification of the German academic system may have discouraged contact between the two men.

5. In the draft of a reply to a graduate student's query in 1970, Gödel indicated that it was precisely his recognition of the contrast between the formal definability of demonstrability and the formal *un*definability of truth that led to his discovery of incompleteness. That he did not bring this out in (1931) is perhaps explained by his observation (in a crossed-out passage from that same draft) that 'in consequence of the philosophical prejudices of [those] times ... a concept of objective mathematical truth ... was received with greatest suspicion and widely rejected as meaningless.' For a fuller discussion of Gödel's avoidance of semantic issues, see Feferman (1985). Re Tarski's theorem on the undefinability of truth, see Tarski (1956), especially the bibliographical note, p. 152; footnote 1, pp. 247-8; the historical note, pp. 277-8; and footnote 2, p. 279. In those notes Tarski makes clear his indebtedness to Gödel's methods, relinquishing, so it would seem, any claim to priority for Gödel's own results (except for the prior exhibition of a consistent yet $\omega$-inconsistent formal system).

6. Church, for example, in a letter to Gödel dated 27 July 1932, stated 'I have been unable to see ... that your conclusions apply to my system' (as, in fact, they did not, since that system was subsequently shown to be inconsistent). In recent correspondence with the author, Church has acknowledged that he 'was among those who thought that the ... incompleteness theorems might be found to depend on the peculiarities of type theory', a theory whose 'unfortunate restrictiveness' he was trying to escape through a 'radically different formulation of logic'. Church goes on to remark, 'indeed [the calculus of $\lambda$-conversion] might be claimed as a system of logic to which the Gödel incompleteness theorem does not apply. To ask in what sense this claim is sound and in what sense not is not altogether pointless, as it may give insight into ... where the boundary lies for applicability of the incompleteness theorem' (A. Church to J. Dawson, 25 July 1983).

7. On his own offprint of Zermelo's remarks, Gödel marked this statement with the annotation 'Dieser Behauptung ist mir ganz nicht verständlich und scheint ... in sich widerspruchsvoll zu sein ....'

8. Quotations and paraphrases of Post in this and the following paragraph are from his letter to Gödel of 30 October 1938.

9. It would be most interesting to know Gödel's opinion of Post's work. I have found no letters to Post in Gödel's *Nachlass* and little mention of Post in Gödel's other correspondence.

10. Not that Hilbert capitulated in the face of Gödel's results; for in the introduction to the first volume of *Grundlagen der Mathematik* (dated March 1934) Hilbert declared:

Im Hinblick auf dieses Ziel [showing the consistency of arithmetic] möchte ich hervorheben, dass die zeitweilig aufgekommene Meinung, aus gewissen neueren Ergebnissen von Gödel folge die Undurchführbarkeit meiner Beweistheorie, als irrtümlich erweisen ist. Jenes Ergebnis zeigt in der Tat auch nur, dass man für die weitergehenden Widerspruchs-freiheitsbeweise den finiten Standpunkt in einer schärferen Weise ausnutzen muss, als dieses bei der Betrachtung der elementaren Formalismen erforderlich ist.

11. I am grateful to the Russell Archives, McMaster University, for permission to quote from this letter.

## REFERENCES

Barzin, Marcel (1940) Sur la portée du théorème de M. Gödel. *Académie Royale de Belgique, Bulletin de la Classe des Sciences*, Series 5, 26: 230-39.

Benacerraf, Paul, and Putnam, Hilary (eds.) (1964) *Philosophy of Mathematics: Selected Readings*. Englewood Cliffs, N.J.: Prentice-Hall.

Church, Alonzo (1946) Review of Finsler (1944). *Journal of Symbolic Logic* 11: 131-2.

Davis, Martin (1965) *The Undecidable*. Hewlett, N.Y: Raven Press.

—— (1982) Why Gödel didn't have Church's Thesis. *Information and Control* 54: 3-24.

Dawson, Jr., John W. (1984) Discussion on the Foundation of Mathematics. *History and Philosophy of Logic* 5: 111-29.

—— (1985) Completing the Gödel-Zermelo correspondence. *Historia Mathematica.* 12: 66-70.

Feferman, Solomon (1985) Conviction and caution: a scientific portrait of Kurt Gödel. *Philosophia Naturalis*; Paul Weingartner and Christine Pühringer (eds.), *Philosophy of Science — History of Science. A Selection of Contributed Papers of the 7th International Congress of Logic, Methodology and Philosophy of Science, Salzburg, 1983*, Meisenheim/Glan, Verlag Anton Hain (1984).

Finsler, Paul (1926) Formale Beweise und die Entscheidbarkeit. *Mathematische Zeitschrift* 25: 676-82. (English translation in van Heijenoort (1967), pp. 440-45.)

—— (1944) Gibt es unentscheidbare Sätze? *Commentarii Mathematici Helvetici* 16: 310-20.

Gödel, Kurt (1930) Einige metamathematische Resultate über Entscheidungsdefinitheit und Widerspruchsfreiheit. *Anzeiger der Akademie der Wissenschaft in Wien* 67: 214-15. (English translation in van Heijenoort (1967), pp. 595-6.)

—— (1930/31) Über Vollständigkeit und Widerspruchsfreiheit. *Ergebnisse eines mathematischen Kolloquiums* 3: 12-13. (English translation in van Heijenoort (1967), pp. 616-17.)

—— (1931) Über formal unentscheidbare Sätze der Principia

Mathematica und verwandter Systeme I. *Monatshefte für Mathematik und Physik* 38: 173-98. (English translation in van Heijenoort (1967), pp. 596-616; reprinted Chapter 2, this volume).

—— (1934) On undecidable propositions of formal mathematical systems. Mimeographed notes by S.C. Kleene and J.B. Rosser of lectures by Kurt Gödel at the Institute for Advanced Study. (As reprinted in Davis (1965), pp. 39-74.)

Gonseth, Ferdinand (ed.) (1941) *Les Entretiens de Zürich sur les fondements et la méthode des sciences mathématiques, 6-9 décembre 1938.* Zürich: Leeman.

Grattan–Guinness, Ivor (1979) In Memoriam Kurt Gödel: His 1931 correspondence with Zermelo on his incompletability theorem. *Historia Mathematica* 6: 294-304.

—— (1981) On the development of logics between the two world wars. *American Mathematical Monthly* 88: 495-509.

Grelling, Kurt (1937/38) Gibt es eine Gödelsche Antinomie? *Theoria* 3: 297-306. *Zusätze und Berichtigungen,* 4: 68-9.

Hahn, Hans, *et al.* (1931) Diskussion zur Grundlegung der Mathematik. *Erkenntnis* 2: 135-51. (English translation in Dawson (1984), pp. 116-28.)

Helmer, Olaf (1937) Perelman *versus* Gödel. *Mind* 46: 58-60.

Kleene, Stephen C. (1937a) Review of Perelman (1936). *Journal of Symbolic Logic* 2: 40-41.

—— (1937b) Review of Helmer (1937). *Journal of Symbolic Logic* 2: 48-9.

—— (1976/78) The Work of Kurt Gödel. *Journal of Symbolic Logic* 41: 761-78. Addendum, 43: 613.

—— (1981a) Origins of recursive function theory. *Annals of the History of Computing* 3: 52-67. (Corrigenda in Davis (1982), footnotes 10 and 12.)

—— (1981b) The theory of recursive functions, approaching its centennial. *Bulletin of the American Mathematical Society (n.s.)* 5:1, 43-61.

Kreisel, Georg (1979) Review of Kleene (1978). (Addendum to Kleene (1976).) *Zentralblatt für Mathematik und ihre Grenzgebiete* 366: 03001.

Kuczyński, Jerzy (1938) O Twierdzeniu Gödla. *Kwartalnik Filozoficzny* 14: 74-80.

Ladrière, Jean (1957) *Les Limitations Internes des Formalismes.* Louvain: E. Nauwelaerte. Paris: Gauthier-Villars.

Menger, Karl (1978) *Selected Papers in Logic and Foundations, Didactics, Economics.* (Vienna Circle Collection, Number 10.) D. Reidel: Dordrecht-Boston-London.

Mostowski, Andrzej (1938) Review of Kuczyński (1938). *Journal of Symbolic Logic* 3: 118.

Perelman, Charles (1936) L'Antinomie de M. Gödel. *Académie Royale de Belgique, Bulletin de la Classe des Sciences,* Series 5, 22: 730-36.

Popper, Karl R. (1980) Der wichtigste Beitrag seit Aristoteles. *Wissenschaft aktuell* 4/80: 50-51.

Post, Emil L. (1965) Absolutely unsolvable problems and relatively undecidable propositions: account of an anticipation. In Davis (1965), pp. 338-433.

Quine, Willard V. (1978) Kurt Gödel, (1906-1978). *Year Book of the American Philosophical Society* 1978: 81-84.

Rosser, J. Barkley (1936) Extensions of some theorems of Gödel and Church. *Journal of Symbolic Logic* 1: 87-91. (Reprinted in Davis (1965), pp. 231-35.)

—— (1938) Review of Grelling (1937/38). *Journal of Symbolic Logic* 3: 86.

Tarski, Alfred (1956) *Logic, Semantics, Metamathematics.* Edited and translated by J.H. Woodger. Oxford: Oxford University Press.

Turing, Alan M. (1936/37) On computable numbers, with an application to the Entscheidungsproblem. *Proceedings of the London Mathematical Society*, Series 2, 42: 230-65. Corrigenda, 43: 544-46. (Reprinted in Davis (1965), pp. 115-54.)

van Heijenoort, Jean (ed.) (1967) *From Frege to Gödel. A Source Book in Mathematical Logic, 1879-1931.* Cambridge, Mass.: Harvard University Press.

Wang, Hao (1981) Some facts about Kurt Gödel. *Journal of Symbolic Logic* 46: 653-9.

Zermelo, Ernst (1932) Über Stufen der Quantifikation und die Logik des Unendlichen. *Jahresbericht der deutschen Mathematiker-Vereinigung* 41, part 2: 85-8.

# V

# Kurt Gödel: Conviction and Caution

Solomon Feferman

In the course of preparing an introductory chapter on Gödel for a forthcoming comprehensive edition of his works,[1] I was struck by the great contrast between the deep platonist convictions Gödel held concerning the objective basis of mathematics and the special caution he exercised in revealing these convictions. Gödel says that he had arrived at a position of philosophical realism early in his university years, and he credits his enormous successes in mathematical logic during the 1930s almost entirely to his holding this point of view.[2] Yet there is hardly a word throughout that period giving any indication of his attitude; indeed the first open expression of it came only in 1944. I was led to seek the reasons for his guarded disposition and to speculate on whether it might have affected his work and choice of problems, especially whether there were things that he *refrained* from doing in consequence. In particular, it seemed to me that he could well have been more centrally involved in the development of the fundamental concepts of modern logic — *truth* and *computability* — than he was; in fact his role turned out to be peripheral in both cases.

What follows, then, is a partly speculative essay on these aspects of Gödel's scientific personality. It is intended to be largely complementary to the more settled and generally accepted picture to be offered in the piece mentioned above. However, there is some overlap in the data to be interpreted. For the most part, in considering the above questions I drew on Gödel's own publications and published remarks. As it turned out, the views I developed in the process were reinforced by material found in Gödel's *Nachlass* and not previously available.

It seems to me that the questions raised here are of interest to consider both for the purposes of scientific biography (in this case, of Gödel) as well as for case studies of conceptual analysis (here, of truth and computability).

To provide the needed reference points, it is necessary to begin with a quick survey of Gödel's work. At present, the most complete and handiest source for the interested reader is the annotated bibliography of Gödel's work provided by Dawson (1983).[3] Since not all of Gödel's papers are relevant to the present purposes, I use two systems of reference in the following. Dates given in brackets are used for papers by Gödel listed in Dawson (1983) when they are not the subject of extended commentary here. All other references, with dates in parentheses, are to be found in the bibliography below.

Gödel's work falls naturally into two parts, with 1940 as the dividing line. This also separates the loci of his activities, namely Vienna prior to 1940 and Princeton thereafter. Gödel entered the University of Vienna as a student in 1924, finishing with a doctorate in 1929. He continued to be based in Vienna during the period 1929-1939, becoming *Privatdozent* at the university in 1931. But he also made three academic visits to the United States during this time: twice to Princeton and once to Notre Dame. In 1940, Gödel became a member of the Institute of Advanced Study (IAS) in Princeton, where he eventually became a professor; he remained there until his death in 1978.

The period to 1940 contains Gödel's most famous contributions to mathematical logic: the completeness theorem (1930), obtained in his dissertation (1929), the incompleteness theorems (1931) and the consistency of the axiom of choice and the (generalized) continuum hypothesis (1938-40). There are further less well-known but still important results from this period: on the decision problem, intuitionistic logic and arithmetic, speed-up theorems and even the propositional calculus. Also of interest are some notes on geometry, most of which had been overlooked until rediscovered by Dawson (1983).

In the early 1940s, Gödel continued to work on problems of mathematical logic. He found a new quantifier-free functional interpretation of intuitionistic logic as early as 1941, though the results were not published until (1958) in the journal *Dialectica* (it has thus been dubbed 'the Dialectica interpretation').[4] He also grappled for some years with the problem of the independence of the axiom of choice and the continuum hypothesis from the axioms of set theory, attaining only partial success with the first of these problems (never published). Increasingly Gödel turned his attention to the philosophy of mathematics, represented by several publications beginning with (1944). The significance of this philosophy for work in set theory was brought out in the expository paper (1947) on the continuum problem. During the latter part of the 1940s, Gödel made an unusual contribution to relativistic cosmology which had significance for the philosophy of space and time. From Wang (1981) and other sources we know that Gödel was preoccupied with problems of general philosophy and metaphysics in the 1950s and continuing on thereafter.

The article (1944) was Gödel's contribution to the volume of the *Living Philosophers* series which was devoted to Bertrand Russell. He provided there a critique of Russell's mathematical logic both from a technical standpoint and with respect to the underlying interpretation of the theory of types. Russell had held a 'no class' interpretation: this led him to adopt a ramified theory, though the *prima facie* problems of developing analysis in such a theory impelled Russell to adjoin the *ad hoc* Axiom of Reducibility. The latter move was criticized (in the 1920s) by Ramsey, who pointed out that doing so nullified the effect of ramification and that the whole could be just as well replaced by the simple theory of types. But it was Gödel who emphasized in (1944) that one must look to the conception of classes as entities having an existence independent of human thoughts and constructions in order to provide the proper interpretation of the resulting theory (as well as various theories of sets). Perhaps to ward off criticisms of this strong platonist (philosophically realist) position, Gödel compared the assumption of the existence of underlying mathematical objects with that of underlying physical objects, arguing that such assumptions were needed in both cases to deduce the data of ordinary experience, and were necessary to obtain a satisfactory account of that experience. Incidentally, Gödel mentioned in (1944) that a transfinite extension of the ramified hierarchy had proved useful in his own work, by which he meant the employment of it in (1938-40) to define the constructible hierarchy of sets and to obtain a model of the axiom of choice and the (generalized) continuum hypothesis.

In (1947) Gödel stated his philosophical viewpoint for the framework of Zermelo–Fraenkel set theory ZF, in terms of its informal interpretation in the cumulative hierarchy of sets. This is obtained by transfinite iteration of the power set operation, where at each stage the set of all subsets (or power set) of a given set is supposed to exist independently of human constructions. Gödel says that Cantor's notion of (infinite) cardinal number is definite and unique. In particular, the set of all subsets of $\omega$ has cardinal $2^{\aleph_0}$, which is the same as the cardinal of the continuum. Since the axiom of choice AC must be granted to be true when there are no limitations on set construction, each set must be in one–one correspondence with a definite aleph, $\aleph_\alpha$. Thus the statement $2^{\aleph_0} = \aleph_1$ of the continuum hypothesis CH has a determinate truth value. The fact that efforts to prove it or disprove it had previously failed was neither here nor there, according to Gödel. Though he had himself proved CH (and GCH: $2^{\aleph_\alpha} = \aleph_{\alpha+1}$) consistent with ZFC (= ZF + AC), this told us nothing concerning its truth value. Indeed Gödel believed that CH was false, since it had (according to him) counter-intuitive consequences. Thus, since he took the axioms of ZFC to be true, he anticipated that CH would be independent of ZFC. This was finally established by Cohen in 1963, just in time for Gödel to be able to include that information in a postscript to his revision [1964a] of (1947). Though Gödel may have felt bolstered in his view of CH by this outcome, the question of truth of CH was still unsettled by it. Gödel himself thought that the continuum problem might eventually be settled by adjoining strong axioms of infinity; however all subsequent work on plausible extensions of ZFC by such axioms has left CH undecided.

Further published expressions of Gödel's platonist philosophical views are to be found in his Princeton Bicentennial remarks of 1946 (first appearing in [1965d]), and in the letters quoted in Wang (1974). While set-theoretic notions receive the most attention, a reading of Gödel (1958) and, again of (1944), would suggest that his view of the objective character of fundamental mathematical entities and the definiteness of questions concerning them seems to go beyond set-theoretical realism to comprehend also such notions as abstract concepts (1944) and constructive functions and proofs (1958).

In the two letters to Wang dated 7 December 1967 and 7 March 1968, quoted on pp. 8-11 of Wang (1974), Gödel tells us in the strongest terms that his philosophical views played an essential role in his work from the very beginning (1929).

According to Gödel, though Skolem already had all the machinery needed for the completeness theorem for the 1st order predicate calculus in 1922, he did not arrive at the theorem itself because he lacked the 'required epistemological attitude toward metamathematics and toward non-finitary reasoning'.[5] Gödel goes on in these letters to stress, similarly, the importance of his 'objectivistic conception of mathematics and metamathematics in general, and of transfinite reasoning in particular' for his further great successes of (1931) and (1938-40). We shall elaborate on those passages below.

It is clear from the passage above that Gödel held his 'objectivistic' views by the time of his 1929 dissertation. Unpublished material found in the *Nachlass* allows us to push that date back even further. It happens that Burke D. Grandjean, an Instructor in Sociology at the University of Texas (Austin) wrote Gödel several times in 1974-75, seeking information on Gödel's background in connection with research he was doing on the social and intellectual situation in central Europe during the first third of the 20th century. For this purpose he also sent Gödel an individually designed questionnaire. Found in the files is a typed response from Gödel dated 19 August 1975, together with the questionnaire which was dutifully filled out, though neither was ever sent to Grandjean![6] In his letter, Gödel refers to his (1964 revision of the) paper (1947) and to the quotations from Wang (1974) for a representation of his philosophical views, which he terms a form of conceptual and mathematical realism; he then goes on to say that *he held these views since about 1925, in other words early in his university years.* But Gödel relates that his interest in philosophy dates back even farther (along with mathematics) to around 1921-22; in particular, he first studied Kant at that time. There is much in the letter and questionnaire concerning his relationship with the 'Vienna Circle' and the standpoint of logical positivism (or empiricism) to which he was directly opposed philosophically. This supports what we know from other sources, and is a matter to which we shall return below (cf. note 18).

A general question may be raised about how much to accept of these retrospective reports, in which Gödel speaks of his state of mind and attitudes some forty or fifty years previously. However, there is no contrary evidence that I know of which would throw doubt on them, and all the information we have available makes a rather coherent picture when assembled. Although it would be a mistake to assume that Gödel's sophisticated views in (1944) and (1947) were already worked out in his university days, it is safe to say that Gödel already had firm general platonistic views by the time he came in contact with the Vienna Circle.

Gödel was introduced to the Circle, which centered around Moritz Schlick, by his teacher Hans Hahn. This was around 1926; he attended meetings fairly regularly during 1926-28 and after that gradually moved away from the Circle. From Wang (1974), pp. 7-13, Wang (1981), p. 653 and the 1975 letter and questionnaire for Grandjean mentioned above, we know that Gödel disagreed with their fundamental tenet concerning the nature of mathematics according to which it was true by convention as the 'syntax of language'. Nevertheless, he did not make his opposition to these views openly known at the time. Gödel grants that his interest in foundational problems was influenced by the Vienna Circle, considering the prominence it gave to logic and the logicist programme of Russell and Whitehead in their *Principia Mathematica*.

For the specific problems in mathematical logic and the foundations of mathematics which Gödel came to pursue, the figures who loomed in his mind would no doubt have been Hilbert and Brouwer. Both had vigorously promoted foundational schemes opposed to the platonism implicit in Cantorian set theory. Hilbert's views had varied over the years, but always took a formalist direction, settling in the 1920s into his programme to obtain finitist consistency proofs for formal mathematical systems. Earlier he had taken a position equating existence of mathematical concepts with the consistency of axiom systems provided for them. While this was not a necessary part of Hilbert's finitist programme, the emphasis on consistency instead of correctness (according to an informal interpretation) carried a residue of that position. Brouwer on the other hand rejected non-constructive existence proofs and the Cantorian concepts of 'actual' infinities altogether, and attempted to rebuild mathematics on competely constructivist grounds, using his own intuitionistic version of constructivism.

In his thesis (1929),[7] Gödel solved the problem concerning the completeness of an axiom system for the first-order predicate calculus which had been posed by Hilbert and Ackermann in their 1928 book on mathematical logic. He showed that if an axiom system is consistent then it has a model; Gödel's proof used König's lemma as an essential non-constructive step. In a sense, Gödel's result could be said to justify the equation between existence and consistency. Gödel takes pains in his introduction to argue against this as an *a priori* philosophical claim (that the criterion for existence of a mathematical concept is simply the consistency of a system for it). In this respect he implicitly criticized Hilbert, though not identifying the position itself with Hilbert.[8] In other words, he is concerned to explain on this flank why his result is needed at all. Against Brouwer, on the other hand, he first wards off any attempt to interpret the completeness result itself in constructive terms, pointing out that this would lead to the decidability of the predicate calculus.[9] In addition, he says that there is no reason to restrict oneself to constructive methods of proof, since the problem dealt with is a mathematical one like any other.

What is of interest for the present essay is that these incisive (and pregnant) remarks were completely removed when Gödel came to publish his (1930) version of the thesis. Whether this was at Hahn's urging or on his own initiative is not known; in the latter case it would be the first evidence of the caution which is of concern here. But his retrospective comments suggest that in 1929-30 he already saw himself as being opposed to the dominant viewpoint of the times. In his letter to Wang of 1967 (Wang (1974), pp. 8-9), Gödel speaks of the 'blindness (or prejudice, or whatever you may call it) of logicians' at that time, according to which non-finitary reasoning was not accepted as a meaningful part of metamathematics. 'But now the aforementioned easy inference from Skolem 1922 is definitely non-finitary, and so is any other completeness proof for the predicate calculus. Therefore these things escaped notice or were disregarded.'[10]

Gödel did make one public statement of a philosophical character in the period directly following his thesis work. This was at an important symposium on the foundations of mathematics held in Königsberg in 1930, at which Carnap, Heyting, von Neumann and Waismann presented the positions of logicism, intuitionism, formalism and logical empiricism respectively. In the ensuing discussion Gödel openly criticized the view which took consistency as the criterion of existence. He then went on to announce publicly for the first time the existence of undecidable propositions: '[Assuming the consistency

of classical mathematics] one can even give examples of propositions ... which are really contentually true [*inhaltlich richtig*]
but are unprovable in the formal system of classical mathematics.' Hence, the negations of such statements could be
adjoined to form a consistent system containing false statements; thus consistency would not guarantee existence in the
intended sense. These remarks appear in [1931a], by which
time the famous (1931) paper had appeared. It seems that the
significance of Gödel's remarks was appreciated only by a few
participants (von Neumann among them) at the meeting itself.
For a translation of the full discussion, with background and
commentary see Dawson (forthcoming).

This now brings us to the central topic here, namely the role
of the notion of truth in Gödel's incompleteness results (and,
later, in his development of the constructible hierarchy). Let me
begin by assembling what Gödel himself said on the subject in
his publications and then what he added in other communications.

The paper (1931) commences with a sketch of the proof of
the first incompleteness theorem: for a specified formal system
PM (adapted from *Principia Mathematica*), the device of
(Gödel-) numbering is used to construct a statement $[R(q);q]$
equivalent to $\overline{\text{Bew}} \, [R(q);q]$, which we may interpret as
expressing of itself that it is not provable in PM. Then if it is
provable it would be true, hence by its interpretation it would
not be provable. Thus it is not provable after all, so its negation
is not true, hence not provable. Gödel says: 'The analogy of this
argument with the Richard antinomy leaps to the eye. It is
closely related to the "Liar" too.'[11] He goes on to say:

> The method of proof just explained can clearly be applied to
> any formal system that, first, when interpreted as repre-
> senting a system of notions and propositions, has at its dis-
> posal sufficient means of expression to define the notions
> occurring in the argument above (in particular, the notion of
> 'provable formula') and in which, second, every provable
> formula is true in the interpretation considered. The purpose
> of carrying out the above proof with full precision in what
> follows is, among other things, to replace the second of the
> assumptions just mentioned by a purely formal and much
> weaker one.

What this comes down to in the body of the paper is: *consistency* of the system implies non-provability of $[R(q);q]$, and ω-*consistency* implies non-provability of its negation. Evidently both hypotheses are weak consequences of *truth* of the system in a given interpretation.

Gödel lectured on his incompleteness results at the Institute for Advanced Study in the spring of 1934, during his visit for the academic year 1933-34. Notes of the lectures were taken by S.C. Kleene and J.B. Rosser, and after they were reviewed and corrected by Gödel, mimeographed copies were made and circulated fairly widely. But they did not appear in print until they were included in Davis's 1965 collection *The Undecidable*; Gödel's notes are thus listed as [1965a] in Dawson (1983), though we should of course ascribe 1934 as the basic date of appearance. In Section 7 of these notes, Gödel discusses the relation of his arguments to the paradoxes, in particular that of Epimenides ('The Liar'). He argues that for a person A to say (on a given day) that every statement he makes (on that day) is false — can be made precise only if A specifies a language B and says that '... every statement that he made in the given time was a false statement in B. But "false statement in B" cannot be expressed in B, and so his statement was in some other language, and the paradox disappears.' In other words, 'the paradox can be considered as a proof that "false statement in B" cannot be expressed in B.' To this Gödel appends the following footnote appearing in [1965a], but not in the original 1934 version of the notes:

> For a closer examination of this fact see A. Tarski's papers published in: *Trav. Soc. Sci. Lettr. de Varsovie*, Cl. III No. 34, 1933 (Polish) (translated in: *Logic, Semantics Metamathematics, Papers from 1923 to 1938 by A. Tarski*, see in particular p. 247 ff.) and in: Philosophy and Phenom. Res. 4 (1944), p. 341-376. In these two papers the concept of truth relating to sentences of a language is discussed systematically.

(There is also a reference to Carnap in this footnote.)

In the introduction by A.W. Burks to von Neumann (1966), edited from lectures and MSS of von Neumann on automata, Burks refers to correspondence which he had with Gödel on one passage that had puzzled him. Gödel's response is quoted von Neumann (1966), pp. 55-6:

I think the theorem of mine which von Neumann refers to is not that on the existence of undecidable propositions or that on the lengths of proofs but rather the fact that a complete epistemological description of a language A cannot be given in the same language A, because the concept of truth of sentences of A cannot be defined in A. It is this theorem which is the true reason for the existence of undecidable propositions in the formal systems containing arithmetic. I did not, however, formulate it explicitly in my paper of 1931 but only in my Princeton lectures of 1934. The same theorem was proved by Tarski in his paper on the concept of truth published in 1933 in *Act. Soc. Sci. Lit. Vars.*, translated on pp. 152-278 of *Logic, Semantics and Metamathematics.*

Let us look next at Gödel's report of how he arrived at the incompleteness results in the first place, as reported by Wang in (1981) p. 654. After his thesis, Gödel began studying Hilbert's problem to prove the consistency of analysis by finitist means. Finding this restriction on methods of proof 'mysterious' and aiming to 'divide the difficulties', '... his idea was to prove the consistency of number theory by finitist number theory, and prove the consistency of analysis by number theory, where one can assume the truth of number theory, not only the consistency'. In other words, a relative consistency proof was sought of analysis to number theory. Wang (1981) goes on to say:

> [Gödel] represented real numbers by formulas ... of number theory and found he had to use the concept of truth for sentences in number theory in order to verify the comprehension axiom for analysis. He quickly ran into the paradoxes (in particular, the Liar and Richard's) connected with truth and definability. He realized that truth in number theory cannot be defined in number theory and therefore his plan ... did not work.

But Gödel was then able to go on to realize the existence of undecidable propositions in suitably strong systems.

We can reconstruct and flesh out this account as follows. If one were to try to give a relative consistency proof of analysis (in its form as second order number theory) in first order number theory by a formal model or interpretation, the obvious idea would be to interpret the set variables as ranging over the arithmetically definable sets. Equivalently, we could number formulas of arithmetic with one free variable $x$, say $A_n(x)$ ($n = 0, 1, 2, \ldots$) and interpret the set variables as ranging over $\omega$, with $m \in n$ interpreted as: $A_n(m)$ is true. But for this model (i.e. to interpret $x \in y$ as a formula of arithmetic with two free variables) one needs the *general notion of truth* of sentences of number theory. This would be problematic in view of the classical paradoxes. Gödel's key step was to realize that definite sense could be given to the phrase 'this statement' in the formulation of the Liar paradox, by a diagonal substitution construction carried out in arithmetic. While one might question whether 'this statement is false' is indeed a statement of natural language, there was no question that by Gödel's argument one could construct a statement of number theory with number $q$ having specified self-referential properties. Thus if truth for number theory *were* definable within itself, one could find a precise version of the Liar statement, giving a contradiction. It follows that truth is not so definable. But provability in the system *is* definable, so the notions of provability and truth must be distinct. In particular, if all provable sentences are true, there must be true non-provable sentences. The self-referential construction applied to provability (which *is* definable) instead of truth, then leads to a specific example of an undecidable sentence. The technical work of (1931) goes into: giving precise sense to the notion of definability within a formal system, verifying that provability in the system is so definable, carrying out the diagonal substitution construction, and checking that the hypotheses of consistency, resp. $\omega$-consistency suffice to carry through the argument.[12]

We may conclude from all this that the concept of truth in arithmetic was for Gödel a definite objective notion, and that he had arrived at the undefinability of that notion in arithmetic by 1931. On the other hand, he did not state this as a result (only done so later, and independently, by Tarski in 1933), and he took pains to eliminate the concept of truth from the main results of 1931. This raises a series of questions, the first being: why? Gödel's own answer is contained in part in the correspondence reproduced in Wang (1974), p. 9. Following the passages quoted above on the importance of his philosophical views for arriving at the completeness theorem, Gödel says:

I may add that my objectivist conception of mathematics and

metamathematics in general, and of transfinite reasoning in particular, was fundamental also to my other work in logic. How indeed could one think of *expressing* metamathematics *in* the mathematical systems themselves, if the latter are considered to consist of meaningless symbols which acquire some substitute of meaning only *through* metamathematics ... it should be noted that the heuristic principle of my construction of undecidable number theoretical propositions in the formal systems of mathematics is the highly transfinite concept of 'objective mathematical truth' as *opposed* to that of 'demonstrability' (cf. M. Davis, *The Undecidable*, New York 1965, p. 64 where I explain the heuristic argument by which I arrive at the incompleteness results), with which it was generally confused before my own and Tarski's work.

Later he adds (p. 10), '... formalists considered formal demonstrability to be an *analysis* of the concept of mathematical truth and, therefore were of course not in a position to *distinguish* the two.' This explains Gödel's view of why he had succeeded where others had failed (the general theme of the two letters quoted in Wang (1974), pp. 8-11) but only indirectly why he eliminated the concept of truth from his work.

A more direct explanation is found in a reply to a letter dated May 27, 1970 from a graduate student named Y. Balas. The reply, which is in Gödel's *Nachlass* in draft form, is unsigned and is marked 'nicht abgeschickt' (cf. note 5). Gödel explains here how he arrived at the undecidable propositions, including his attempt to give a 'relative model-theoretic consistency proof of analysis in arithmetic ... [by use of] an arithmetical $\epsilon$-relation satisfying the comprehension axiom.' He goes on to criticize Finsler's earlier attempt to prove 'formal undecidability in an absolute sense', which he terms a 'nonsensical aim'.[13] But of greatest interest for the question raised here is the following vehement paragraph, which was crossed out in the draft reply.

However in consequence of the philosophical prejudices of our times 1. nobody was looking for a relative consistency proof because [it] was considered axiomatic that a consistency proof must be finitary in order to make sense 2. a concept of objective mathematical truth as opposed to demonstrability was viewed with greatest suspicion and widely rejected as meaningless.

Here, in a crossed-out passage in an unsent reply to an unknown graduate student, I think we have reached the heart of the matter. Despite his deep convictions as to the objectivity of the concept of mathematical truth, Gödel feared that work assuming such a concept would be rejected by the foundational establishment, dominated as it was by Hilbert's ideas. Thus he sought to extract results from it which would make perfectly good sense even to those who eschewed all non-finitary methods in mathematics. In doing so, he got a pay-off which apparently even he did not anticipate,[14] namely the second incompleteness theorem — according to which no sufficiently strong consistent formal system can prove its own consistency. Even more, once Gödel realized the generality of his incompleteness results it was natural that he should seek to attract attention by formulating them for the strong theories that had been very much in the public eye: theories of types such as PM and theories of sets such as ZF (Zermelo–Fraenkel).[15] But if the concept of objective mathematical truth would be rejected in the case of arithmetic, should not one expect an even greater negative reaction in the case of theories of types or sets? All the more reason, then, not to have any result depend on it, and no need then to express one's convictions about it.

What is clear so far is that Gödel anticipated Tarski at least in establishing the undefinability of arithmetic truth within arithmetic. No evidence has been presented, and none is available, to show that Gödel considered the problem which Tarski took to be the main one that he (Tarski) had solved: to give a set-theoretic definition of the notion of truth for the first order language of any structure. Indeed, one may ask whether Gödel thought such a definition would even be necessary; this question should be raised with respect to each of his major results in the 1930s.

To begin with, in his completeness result (1929, 1930), the notion of *validity* of a formula in the first order predicate calculus is understood in an informal sense. Nowadays, *that* is defined in terms of *satisfaction* in every structure (or interpretation) and that in turn is preceded by Tarski's inductive definition of satisfaction. Of course there was a long tradition of use of the informal notion of satisfiability (through the work of Löwenheim, Skolem and others). It may be regarded as being well enough understood at the time of the 1928 book by Hilbert and Ackermann so as not to be considered problematic.[16] At any rate, Gödel, with his objectivist conception of truth, would not have thought it so; but would he have considered it a point on which he could be challenged?

Next there is an interesting footnote 48[a] (evidently an afterthought) in Gödel (1931):

> As will be shown in Part II of this paper, the true reason for the incompleteness inherent in all formal systems of mathematics is that the formation of ever higher types can be continued into the transfinite .... For it can be shown that the undecidable propositions constructed here become decidable whenever appropriate higher types are added .... An analogous situation prevails for the axiom system of set theory.

Since Tarski's work we analyse this by saying that the inductive definition of truth for a language can be given in an extension of that language to a next higher type: the satisfaction relation being the smallest relation satisfying such and such closure conditions. Formally, within a system of set theory $S$ we can prove the consistency of any fragment $S_0$ having a natural model in $V_\alpha$ (the sets of rank $< \alpha$) by defining truth in $(V_\alpha, \in \upharpoonright V_\alpha)$ set-theoretically and proving that all axioms of $S_0$ are true in $(V_\alpha, \in \upharpoonright V_\alpha)$. Since Gödel did *not* write Part II to (1931) and never commented further on footnote 48[a], we do not know whether he saw it necessary to give a set-theoretic analysis of the concept of truth in order to justify his claim. (The significance of footnote 48[a] in this respect was brought to my attention by S. Kripke.)

Finally, there is the implicit use of the set-theoretic definition of truth in (1938, 1939). The constructible hierarchy $L_\alpha$ ($\alpha$ an arbitrary ordinal) is defined inductively, taking $L_{\alpha+1}$ to be the set of all subsets of $L_\alpha$ which are definable in $(L_\alpha, \in \upharpoonright L_\alpha)$ (allowing parameters in $L_\alpha$). But the precise notion of *definability* here requires prior explanation of satisfaction in $(L_\alpha, \in \upharpoonright L_\alpha)$. Moreover, to carry through the absoluteness argument outlined in (1938, 1939), one would have to formalize the definition of truth in a set-theoretical structure $(a, \in \upharpoonright a)$ *within* ZF. Again, Gödel says nothing explicit about the matter. What we do know is that he found in (1940) a way to avoid considering it at all, by working within the Bernays–Gödel system of sets and classes, and replacing all work with formulas by operations on classes. This suggests that he was well aware of what would be needed were he to carry out in detail the proof sketched in (1938, 1939).

These questions as to whether Gödel saw the need for a truth-definition may reward further pursuit; I hesitate to suggest answers on the basis of the present evidence. Only the slim statement in the footnote on p. 64 in Davis (1965) (*added in 1965*) quoted above — where Gödel credits Tarski for the systematic study of the concept of truth — gives any hint of Gödel's view of the matter.

Limitations of space do not permit us to go into a comparable discussion of Gödel's work bearing on the notion of effective computability. In any case, there is much material on this in the very interesting recent historical papers Kleene (1981) and Davis (1984). Basically, Gödel was unconvinced by Church's thesis, since the proposed identification of the effectively computable with the $\lambda$-definable functions did not rest on a direct conceptual analysis of the notion of finite algorithmic procedure. For the same reason he resisted identifying the latter with the general recursive functions in the sense of Herbrand as modified by Gödel. Indeed, at the time (in 1934) that Church was beginning to propound his thesis, Gödel in his Princeton lectures was saying that the notion of effectively computable function could only serve as a heuristic guide. It was only when Turing offered in 1937 the definition in terms of his 'machines' that Gödel was ready to accept such an identification, and thereafter he always referred to Turing's work as having provided the 'precise and unquestionably adequate definition of formal system' by his 'analysis of the concept of "mechanical procedure" ... needed to give a fully general formulation of the incompleteness results'. It is perhaps ironical that the various classes of functions ($\lambda$-definable, general recursive, Turing computable) were proved in short order to be identical, but Gödel cannot be faulted for his reservations on philosphical grounds at that point. Nevertheless, one may ask why Gödel did not pursue such an analysis *himself*.[17] Surely, given the depth of his understanding and experience, he was in a position to do that in 1934 or even earlier. My guess is that he also feared that no such proposal could be made convincing to the mathematical public of his day, just as the concept of truth would not be taken seriously. If so, the subsequent development showed Gödel to have been mistaken. Though certainly there were controversies both about Tarski's analysis of truth and Turing's thesis, they eventually took their place as accepted cornerstones of mathematical logic.

What I have termed Gödel's caution is in many respects understandable. At least initially, from the standpoint of a relative unknown in Vienna, the reaction of the Hilbertian establishment would be of genuine concern. But von Neumann's quick appreciation of (1931) and Gödel's spreading fame should have reassured him that he would not be laughed off stage if he were to go beyond the purely (logico-) mathematical formulation of his results. What I find striking here is the contrast on the one hand between the depth of Gödel's convictions which underlay his work, combined with his sureness of insight leading him to the core of each problem, and on the other hand the tight rein he placed on the expression of his true thoughts. Only when he became established at the Institute and the importance of his fundamental contributions was generally recognized did Gödel finally begin to feel free to let out what he really thought all along.[18] We know that Gödel was very careful in his personal habits, especially as concerned his health, so the caution discussed here is coherent with other characteristics of his personality. Only an in-depth biography could plumb the common sources and establish the interrelationships of the everyday and the scientific personality.

In conclusion, one may wonder how logic might have been different had Gödel been bolder in bringing his philosophical views into play in relation to his logical work, in particular by giving as much importance to concepts as to results. For, throughout the 1930s, he shied away *from new concepts as an object of study* as opposed to *new concepts as a tool for obtaining results* (e.g. the constructible hierarchy). From one perspective, his strategy to avoid controversy was exactly right for the times: his work showed that logic could be pursued mathematically with results as decisive, important and interesting as those from familiar branches of mathematics. But far from leading to endless suspicion and controversy, the fundamental conceptual 'philosophical' contributions of Tarski and Turing added to Gödel's work just what was needed to lay the groundwork for the subject of mathematical logic as we know it today.

111

## NOTES

1. My co-editors on this edition are J.W. Dawson, Jr., S.C. Kleene, G. Moore, R. Solovay and J. van Heijenoort.

2. Sources for these statements will be given below.

3. See also Kreisel (1980), Wang (1981) and Kleene (1984) for more analytic surveys, as well as biographical information.

4. It is of great interest and perhaps relevant to his work on the Dialectica interpretation that Gödel found in the early 1940s some essential errors in Herbrand's arguments for the quantifier-free interpretation which Herbrand had found for the predicate calculus. From material in the *Nachlass* we also know that Gödel sought to correct those errors. This work was never published. The difficulties in Herbrand were eventually rediscovered independently, and the errors corrected a number of years later by Dreben, Andrews and Aanderaa. For more information on this and further references, see van Heijenoort (1967), p. 525.

5. Gödel asserts that Skolem stated a form of the completeness theorem but gave an 'entirely inconclusive argument'. However, it may be questioned whether Skolem even appreciated the completeness theorem in the precise sense first formulated by Hilbert and Ackermann in their 1928 book. In particular, the evidence does not show that Skolem had a definite Hilbert-type formal system in hand for which he stated completeness.

6. The notation 'nicht abgeschicht' is found not infrequently on Gödel's side of his correspondence in the files.

7. One copy of this is available at the University of Vienna; a second (practically identical) copy was found in the *Nachlass*. The dissertation and a translation are to appear in the forthcoming edition of Gödel's works.

8. It is of interest that his criticism is elaborated via an argument, itself inconclusive, which anticipates the possibility of incompleteness results, as eventually obtained in (1931).

9. Undecidability was not established until 1936 by Church, but Gödel must have viewed decidability as unlikely or, at any rate, of another order of difficulty.

10. Gödel says elsewhere that he did not know Skolem's 1922 paper at the time of his thesis work, referring only to an earlier 1920 paper which achieved less.

11. We are using the translation of (1931) in van Heijenoort (1967), pp. 596-616.

12. As we know, Rosser later showed that consistency alone was sufficient to obtain incompleteness on both sides.

13. Cf. van Heijenoort (1967), pp. 438-40.

14. Cf. Wang (1981), pp. 654-5.

15. Cf. the first paragraph of Gödel (1931).

16. There is evidence that Hilbert may have considered it problematic: in his 1928 Bologna speech (ref. *1928a* in van Heijenoort (1967)), Hilbert seeks (in a mistaken way) to replace the semantic notion of satisfiability by a syntactic one. It would be interesting to settle the

respective roles of Hilbert and Ackermann in the writing of their book.
17. This question is raised by Davis (1984) but I think not fully answered there. The reasons for my reservations must however, be put off to another occasion.
18. The following is curious in this respect. Gödel never did state publicly (until his communications to Wang reported in 1974) what he thought of the views of the Vienna Circle. Actually, he was supposed to contribute an article for the volume of the *Library of living philosophers* dedicated to R. Carnap. This volume eventually appeared in 1963 without Gödel's article, despite repeated attempts by the editor of the volume (P.A. Schilpp) to extract it from him. Six versions of this article, entitled 'Is mathematics syntax of language?' have been found in the *Nachlass*; Gödel's answer to the question in his title is, of course, NO.
Hilbert died in 1943, the year before Gödel (1944) appeared.

## REFERENCES

Davis, M. (ed.) (1965) *The Undecidable.* Raven Press, New York.
—— (1984) Why Gödel didn't have Church's thesis. *Information and Control.*
Dawson, J.W., Jr. (1983) The published work of Kurt Gödel: an annotated bibliography. *Notre Dame Journal of Formal Logic* 24, 255-84.
—— (1984) Discussion on the foundation of mathematics (Commentary and translation of *Erkentniss* (1931), 135-51). *History and Philosophy of Logic* 5: 111-29.
Gödel, K. (1929) Über die Vollständigkeit des Logikkalküls. Dissertation, University of Vienna. Reprinted and English translation in Gödel (1986).
—— (1930) Die Vollständigkeit der Axiome des logischen Funktionskalküls. *Monatsh. Math. Phys.* 37, 349-60. Reprinted and English translation in Gödel (1986).
—— (1931) Über formal unentscheidbare Sätze der Principia Mathematica und verwandter Systeme I. *Monatsh. Math. Phys.* 38, 173-98. English translation Chapter 2, this volume.
—— (1938) The consistency of the axiom of choice and the generalized continuum hypothesis. *Proceedings of the National Academy of Sciences of the U.S.A.* 24, 556-7.
—— (1939) Consistency-proof for the generalized continuum-hypothesis. *Proceedings of the National Academy of Sciences of the U.S.A.* 25, 220-24.
—— (1940) The consistency of the axiom of choice and of the generalized continuum-hypothesis with the axioms of set theory. *Annals Math. Studies* 3, Princeton University Press, Princeton (2nd edn., revised with added notes, 1951; 7th printing with added notes, 1966).
—— (1944) Russell's mathematical logic. In *The Philosophy of*

*Bertrand Russell,* Library of Living Philosophers, ed. P.A. Schilpp, Northwestern University Press, Evanston.

—— (1947) What is Cantor's continuum problem? *American Mathematical Monthly* 54, 515-25; errata 55 (1948), 151.

—— (1958) Über eine bisher noch nicht benützte Erweiterung des finiten Standpunktes. *Dialectica* 12, 280-87.

—— (1986) *Collected Works. Volume I:* Publications, 1929-1936, eds. S. Feferman, J.W. Dawson, Jr., S.C. Kleene, G.H. Moore, R.M. Solovay, J. van Heijenoort, Oxford University Press, Oxford.

Kleene, S.C. (1981) Origins of recursive function theory. *Annals of the History of Computing* 3, 52-67.

—— (1985) Kurt Gödel (Biographical Memoir). *Proceedings of the National Academy of Sciences of the U.S.A..*

Kreisel, G. (1980) Kurt Gödel. *Biographical Memoirs, F.R.S.* 26, 148-224.

Tarski, A. (1956) *Logic, Semantics, Metamathematics. Papers from 1923 to 1938,* (ed.) J.H. Woodger, Clarendon Press, Oxford.

van Heijenoort, J. ed. (1967) *From Frege to Gödel, a source book in mathematical logic, 1879-1931.* Harvard University Press, Cambridge, Mass.

von Neumann, J. (1966) *Theory of Self-Reproducing Automata.* ed. A.W. Burks, University of Illinois, Urbana.

Wang, H. (1974) *From Mathematics to Philosophy.* Routledge and Kegan Paul, London.

—— (1981) Some facts about K. Gödel. *Journal of Symbolic Logic* 46, 653-9.

# VI

## On the Philosophical Significance of Consistency Proofs

Michael D. Resnik

Much work in the foundations of mathematics has been motivated by the need to resolve the philosophically disquieting situation created by the discovery of the paradoxes. In particular, consistency proofs for various foundational systems were intended to justify not only the systems themselves but also the retention of the core of classical analysis and set theory. Several of these proofs have been of great mathematical significance, because of the additional side information and the new methods which they furnished. (The best-known example of this is Herbrand's work which led to the completeness theorem for quantification theory and has been used since in studying solvable subcases of the decision problem.) While there is thus no question about the mathematical significance of consistency proofs, their philosophical significance as an answer to scepticism about mathematical theories warrants examination. This is the purpose of the present paper.

It might be remarked that Gödel's theorem about consistency proofs has shown they cannot be used to answer sceptics. However, in recent years work by Feferman has reopened the question. Thus one byproduct of this paper will be to give some publicity to Feferman's results and to call attention to their philosophical importance. In addition, there are the following questions: Given that Gödel's theorem does rule out consistency proofs as an answer to scepticism, (as shall be here argued): (1) How can we answer the sceptic? (2) Why do some consistency proofs still strike as as philosophically important? The conclusion of this paper will attempt to say something sensible about these issues.

## 1. HILBERT'S CONTRIBUTION

In 1900 Hilbert avoided the paradoxes of set theory *via* the axiomatic method. There is no need, he claimed, to give set theoretic constructions of entities such as the real numbers in order to justify their existence. One need only lay down a consistent set of axioms describing them; for the consistency of the axioms is sufficient for the existence of the domain they 'define' and for their truth in it.[1] There are without doubt many problems with this view having to do with the doctrine of implicit definitions,[2] but here let us focus on the role of consistency proofs in Hilbert's doctrine of 1900. Notice first that the doctrine can be updated and even proved as a version of the completeness theorem: every deductively consistent set of sentences has a model. Even so one wonders how this theorem can be applied to a set of axioms $\Gamma$. Obviously, one first needs a consistency proof for $\Gamma$. Frege thought that Hilbert's idea was useless because he knew of no other way to prove consistency except via models.[3] With seventy more years behind us, we know about syntactic or proof theoretic consistency proofs which dispense with models. Indeed, Hilbert invented them, perhaps even in response to Frege. Thus the applications problem has a solution. But there is more to it than that!

Both model theoretic and proof theoretic methods have been used in producing both absolute and relative consistency proofs. (In the latter case one only proves that if $\Gamma$ is consistent so is some $\Gamma'$.) Philosophical attention has been drawn to absolute consistency proofs, however. One has hoped to justify all or at least a portion of mathematics by first formulating it as an axiom system and then finding a proof theoretic consistency proof for these axioms. From an epistemological point of view such proofs are no better than the principles upon which they are based. (This remark holds for relative consistency proofs also.) Thus even if Hilbert's doctrine of 1900 makes sense, it must be qualified; consistency is sufficient for all mathematical truth except for the truth of the principles used in consistency proofs.

Hilbert must have been aware of this limitation because his later doctrine of Finitism presupposes an epistemologically evident basis for proof theoretic consistency proofs. Although it is not clear how Hilbert's finitism is to be formulated, for our present purposes, we can take it to be captured by means of quantifier free primitive recursive arithmetic.[4] Recall that bounded numerical quantification can be defined in terms of finite conjunction and disjunction. Thus the finitistically meaningful sentences (*real* sentences) include identities between terms composed of numerals and symbols for primitive recursive functions and compounds formed from these *via* truthfunctions and bounded quantification. While unbounded quantifications are not counted as real sentences, Hilbert thought that one can get an approximation to universal quantification by construing real sentences, in which one or more numerals are replaced by free variables, as *schemata*. These are meta-linguistic devices for communicating indefinitely many real sentences in one breath.

Although a significant amount of number theory can be developed on this basis, it cannot capture all of classical mathematics. Hilbert's solution to this problem was to assert that all other sentences of mathematics should be construed as meaningless instruments designed for the fruitful manipulation of real sentences. In analogy with the use of ideal elements in geometry — e.g., points in infinity in which parallel lines meet — he called such sentences *ideal* sentences.

Still what if manipulations with ideal sentences lead to false or (worse) contradictory real sentences? Hilbert answered this by placing two conditions on manipulations with ideal sentences. First, they must be confined to the context of finitistically specified formal systems. Roughly this means that the well-formed formulas of the systems must be effectively constructable from a finite alphabet and that we must be able to effectively determine whether a finite sequence of formulas is a proof of its last member. Thus (by using a numerical coding) the formal systems themselves are suitable objects of finitistic investigations. Second, these formal systems must possess finitistic consistency proofs.

Are there real sentences to the effect that a finitistically specified formal system $\Gamma$ is consistent? A natural way to express the consistency of $\Gamma$ is

$$(x)\,(y)\,(z) - (\text{FML}_\Gamma(y).\quad \text{PF}_\Gamma(x, y).\quad \text{PF}_\Gamma(z, \text{neg}(y))).^5(1)$$

However, (1) involves unbounded quantification and is, therefore, finitistically meaningless. But, if we drop the quantifiers from (1) we obtain the finitistically acceptable schema

117

$$- (\text{FML}_\Gamma(y). \quad \text{PF}_\Gamma(x, y). \quad \text{PF}_\Gamma(z, \text{neg}(y))). \tag{2}$$

Any instance of (2) in which '$x$', '$y$', and '$z$' are replaced by expressions in the alphabet of $\Gamma$ will be finitistically significant because '$\text{FML}_\Gamma$', '$\text{PF}_\Gamma$' and 'neg' are effective predicates and operations defined for these expressions. Thus if we can give a finitistic proof (2), we will have an answer to the sceptic with respect to the mathematics formalized in $\Gamma$.

## 2. GÖDEL'S UNDERIVABILITY THEOREMS

Gödel's two famous theorems about formal systems for number theory are usually interpreted as undermining Hilbert's Programme. The first, the incompleteness theorem, has less bearing upon the programme than is often credited to it. If Hilbert had claimed that every sentence of mathematics is true or false and that truth is to be identified with provability in some particular formal system, then the existence of undecidable sentences for systems (otherwise) adequate for mathematics would have undermined his programme. But Hilbert only ascribed literal truth-values to real sentences, and every formal system to which Gödel's theorems apply is complete with respect to its real sentences.[6] Thus the undecidable sentences are ideal sentences. Moreover, since they are far removed from ordinary mathematics there is no compelling reason for a Finitist to include them among his formal theorems. (The argument that there are undecidable but true sentences is not finitistically meaningful, because the sentences involved are ideal.)

Gödel's theorem on consistency proofs has a direct bearing upon Hilbert's Programme. This theorem shows that for a broad class of formal systems $\Gamma$, the 'consistency statements' for $\Gamma$ are unprovable in $\Gamma$, so long as $\Gamma$ is consistent. Thus we might conclude that since the consistency of $\Gamma$ cannot be proved via the methods formalizable in $\Gamma$, it surely cannot be proved via (weaker) finitistic methods alone. Consequently, Hilbert's Programme is destined to fail. Although this conclusion is substantially correct, it will be instructive to examine the situation more closely.

Gödel's results depend upon the establishment of a code between the metalanguage for the formal systems $\Gamma$ in question and the (informal) language of elementary number theory.[7]

Metalinguistic sentences about expressions of $\Gamma$ are encoded as sentences about numbers and their *arithmetical* relations. Metalinguistic predicates encode as arithmetical predicates. Let us italicize a metalinguistic predicate in order to represent its number theoretic encoding. Thus '*FML$_\Gamma x$*' is a number theoretic predicate which is the code for 'FML$_\Gamma x$'. Due to the nature of the code '*FML$_\Gamma x$*' is true of something x just in case x is the code number (Gödel number) of a formula of $\Gamma$.

If $\Gamma$ contains a formalization of intuitive number theory, then the formal language of $\Gamma$ will contain predicates which 'represent' properties and relations of intuitive number theory. Gödel's results concern two particular types of representation called *numerations* and *bi-numerations*, respectively. Let us assume that $\Gamma$ has numerals for each natural number and let ñ be the numeral for the natural number n. Then a predicate **K** of $\Gamma$ *numerates* (in $\Gamma$) a m-ary number theoretic property $K$ just in case: $K(n, \ldots, n_m)$ if and only if $\vdash_\Gamma \mathbf{K}(\bar{n}, \ldots, \bar{n}_m)$ holds for all numbers $n, \ldots n_m$. The predicate **K** *bi-numerates* (in $\Gamma$) $K$ just in case it numerates $K$ and: not $K(n, \ldots, n_m)$ if and only if $\vdash_\Gamma \neg \mathbf{K}(\bar{n}, \ldots \bar{n}_m)$.

Let us return to our metalinguistic example 'FML$_\Gamma x$'. Its Gödel encoding is the number theoretic predicate '*FML$_\Gamma x$*'. However, in the systems under consideration this will have at least one bi-numeration **FML**$_\gamma x$.[8] The same applies to 'PF$_\Gamma$' and 'neg' used above to formulate the metalinguistic consistency statements. With this in mind let us consider Gödel's second theorem.

This theorem states that for a broad class of systems $\Gamma$ and numerations '**PF**$_\gamma$', the sentences

$$(\forall a)\,(\forall b)\,(\forall c) \neg (\mathbf{FML}_\gamma(b) \wedge \mathbf{PF}_\gamma(a, b) \wedge$$
$$\mathbf{PF}_\gamma(c, \mathbf{neg}(b))) \qquad (1')$$

are unprovable in $\Gamma$ if it is consistent. Put loosely, the consistency of $\Gamma$ is unprovable in $\Gamma$ if it is indeed consistent. Yet these are ideal sentences and once again they can be dismissed by a Finitist.

The Finitist would be concerned instead with the 'schema'

$$\neg (\mathbf{FML}_\gamma(p) \wedge \mathbf{PF}_\gamma(n, p) \wedge \mathbf{PF}_\gamma(m, \mathbf{neg}(p))). \qquad (2')$$

in which 'n', 'm', 'p' are 'schematic' for numerals. As $\Gamma$ contains universal generalization the unprovability in $\Gamma$ of (2') follows from the unprovability in $\Gamma$ of (1'). However, all instances of (2') are provable in $\Gamma$, and that this is so is something which can be established finitistically. For (2') is actually a binumeration of a number theoretic predicate of the form $f(n, m, p) = 0$, where $f$ is a recursive function whose value is 1 in case p is the Gödel number of a formula of $\Gamma$, n is the Gödel number of a proof in $\Gamma$ of this formula and m is the Gödel number of a proof of its negation; otherwise $f(n, m, p) = 0$. Thus to prove any instance of (2') we first (effectively) calculate $f(n, m, p)$. If the value is 0 then, letting **f** abbreviate our (bi)-numeration of $f$, we have

$$\vdash_\Gamma \mathbf{f}(\bar{n}, \bar{m}, \bar{p}) = \bar{0}.$$

Indeed, due to the nature of $\Gamma$ and $f$ we can effectively construct a proof in $\Gamma$ of this instance of (2').[9] On the other hand, if $f(n, m, p) = 1$ then we can decode n, m and p and effectively obtain a derivation of a contradiction in $\Gamma$. Once we derive the contradiction in $\Gamma$ we can derive any sentence, in particular the instance of (2') in question. Thus we can in either case construct a proof in $\Gamma$ of the instance.

(*Note*: To carry out the argument it is not necessary to make the finitistically meaningless assumption that $\Gamma$ is or is not consistent. This is avoided by appealing to the finitistically constructed recursive function $f$.)

The result shows that if $\Gamma$ is consistent then the finitist can prove every instance of (2') that he attempts to prove. Each instance also corresponds via the Gödel numbering to a finitistic statement at the metalinguistic level, which, to be sure, is also finitistically provable. However, these merely assert that two sequences of expressions do not constitute proofs of a sentence and its negation; so no finite number of them amounts to a consistency proof for $\Gamma$ (in any sense of consistency). Indeed, only by assuming that $\Gamma$ is consistent can the finitist appeal to the potentially infinite set. Thus our result is of little comfort and we may conclude that Gödel's second theorem does undermine the original Hilbert Programme.

Yet there is something to be gained from this excursion. First, we have already seen that the usual interpretation of Gödel's second theorem is too loose: *there is a sense of consistency in which the consistency of a system can be proved within that system.* Second, the very weakness of this sense of consistency casts doubts upon Hilbert's suggestion that schemata can be used as approximations to unbounded universal quantification.[10]

## 3. FEFERMAN'S CRITIQUE

Hilbert's Programme is so appealing that one might hope to rescue it by modifying the finitistic basis. As doubt has been cast upon Hilbert's views of unbounded quantification we might try liberalizing his programme by allowing unbounded quantification over finitistic entities and extending the principles of finitistic proof to cover the new real statements. With respect to the natural numbers, this would yield the usual elementary number theory. On the other hand, statements of analysis and set theory would remain ideal. This programme is quite attractive, because no one seriously doubts the consistency of elementary number theory while people do have serious doubts about set theory and analysis.[11]

Nonetheless even this extended programme is subject to Gödel's second theorem. For if we cannot prove a sentence of the form (1') in, say, a system $\Gamma$ for set theory, then we cannot prove it using elementary number theory. For if we could carry out such a proof, we could easily obtain a translation of it in $\Gamma$ by applying one of the reductions of arithmetic to set theory. Indeed, the theorem appears to defeat any programme which depends upon the selection of an epistemologically favoured basis for consistency proofs for formal systems.[12]

Before this conclusion can be endorsed Feferman's critique of Gödel's work must be considered. This will require some additional terminology.

Let P be the well known first order formalism for number theory with addition and multiplication. Let us consider extensions of P obtained by adding new function symbols to P together with primitive recursive 'defining axioms' for these symbols. (The induction schema is also extended to cover the additional predicates (atomic and complex) available in the extension.) For example, P can be extended in this way by adding the function symbol 'p(x)' (the predecessor function) and the axioms

$$p(0) = 0$$
$$p(x') = x.$$

Such extensions of P together with their extensions via the same methods are called *primitive recursive extensions* (P. R.) of P. A predicate of P or one of its P. R. extensions P' which is constructed without the use of unbounded quantification is called a *PR-predicate* of P, respectively P'. If **K** is a PR-predicate of P (or P') then $(\exists x_1) \ldots (\exists x_n)$**K** is an *RE-predicate* of P (or P').[13]

The 'consistency sentences' for a system $\Gamma$, having the form (1'), are defined in terms of a 'proof predicate' '$\mathbf{PF}_\gamma$'. This is in turn defined via an axiom predicate '$\gamma$' which is supposed to represent, through numeration or bi-numeration, a number theoretic predicate '$AX_\Gamma$' which is true of all and only the Gödel numbers of the axioms of $\Gamma$. Let '$\mathbf{CON}_\gamma$' abbreviate the sentence defined thus in terms of '$\gamma$'. ($\Gamma$ and P need not be the same; '$\mathbf{CON}_\gamma$' is a sentence of a PR-extension, P', of P.)[14]

Gödel's second theorem shows that a particular 'consistency sentence' for P is not provable in P. Many people believed that this result generalized to all consistency sentences for all systems 'containing' P and often used this assumption uncritically in metamathematical arguments and related philosophical discussions.[15] But Feferman has shown the need for caution in this area. First he stated a precise generalization of Gödel's second underivability theorem. This is the following theorem.

THEOREM 1: Suppose that $\Gamma$ is a consistent extension of P, that $AX_\Gamma$ is the class of Gödel numbers of the axioms of $\Gamma$ and that '$\gamma$' is an RE-numeration of '$AX_\Gamma$' *in some subsystem of $\Gamma$* which is itself an extension of Robinson's system Q.[16] Then '$\mathbf{CON}_\gamma$' is a theorem of neither $\Gamma$ nor P. Gödel's theorem is a special case of this because P contains Robinson's system Q.

Next Feferman showed that there are systems $\Gamma$ in which '$\mathbf{CON}_{\gamma 1}$' *is provable* for appropriate choices of '$\gamma_1$' *which also numerate $AX_\Gamma$*. These systems must be *reflexive*, that is capable of proving the consistency of each of their finite subsystems. More precisely, $\Gamma$ is reflexive just in case $\vdash_\Gamma \mathbf{CON}_\varphi$ for each predicate $\varphi$ of the form

$$x = \bar{n}_1 \lor x = \bar{n}_2 \lor \ldots \lor x = \bar{n}_m,$$

where $n_1$, $n_2$, ..., $n_m$ are Gödel numbers of some of the axioms of $\Gamma$. (P and ZF-set theory are examples of reflexive systems.) Feferman's theorem can be stated as follows:

THEOREM 2: Suppose that $\Gamma$ is a consistent, reflexive extension of P and $AX_\Gamma$ is recursive. Then there is a bi-numeration '$\gamma_1$' of $AX_\Gamma$ in $\Gamma$ such that $\vdash_p \mathbf{CON}_{\gamma 1}$.[17]

Put loosely, this implies that there is a *sense* of consistency in which the consistency of analysis and set theory can be proved in elementary number theory. Perhaps, there is an answer to scepticism after all!

Before celebrating we should at least determine the relationship between this new sense of consistency and the one covered by Theorem 1. At the very least they should be extensionally equivalent; although it is clear that even if they are, this cannot be proved in P or $\Gamma$ (so long as they are consistent). For otherwise, the laws of interchange present in these systems would permit the proof of '$\mathbf{CON}_\gamma$' contrary to Theorem 1.[18] A sketch of Feferman's proof will furnish additional information.

Since $AX_\Gamma$ is recursive, there is a '$\gamma$' which bi-numerates it in P. Define '$\gamma_1$' by

$$\gamma_1 (a) \equiv \gamma(a) \wedge (b) (b \leqslant a \supset \mathbf{CO}_{\gamma \upharpoonright b}),$$

where '$\gamma \upharpoonright b$' abbreviates '$\gamma(x) \wedge x \leqslant b$'. (Thus $AX_\Gamma i_b$ is a finite (possibly empty) set of Gödel numbers of axioms of $\Gamma$). If $AX_\Gamma(n)$ then $\vdash_p \gamma(\check{n})$ (since '$\gamma$' bi-numerates $AX_\Gamma$ in P). But then $\vdash_\Gamma \gamma(\check{n})$. Furthermore, due to the reflexivity of $\Gamma$, $\vdash_\Gamma \mathbf{CON}_{\gamma \upharpoonright 0} \wedge CON_{\gamma \upharpoonright T} \ldots \mathbf{CON}_{\gamma \upharpoonright \Gamma}$. Thus $\vdash_\Gamma \gamma_1 (\check{n})$. On the other hand, if not $AX_\Gamma(n)$, $\vdash_\Gamma \neg \gamma(\check{n})$ and hence $\vdash_\Gamma \neg \gamma_1 (\check{n})$. It follows that '$\gamma_1$' bi-numerates $AX$ in $\Gamma$.

The proof of $\vdash_p \mathbf{CON}_{\gamma 1}$ is obtained by formalizing the following argument in P. (a) Suppose that $\mathbf{CON}_\gamma$. Then all finite subsystems of $\Gamma$ are consistent, so for any x, and y, x $\leqslant$ y only if $\mathbf{CON}_{\gamma \upharpoonright y}$. Hence $\gamma(x)$ just in case $\gamma_1 (x)$. So $\mathbf{CON}_{\gamma 1}$. (b) Suppose that $\neg \mathbf{CON}_\gamma$. Then some finite subset of $\Gamma$ is inconsistent, i.e., $(\exists x) (\neg \mathbf{CON}_{\gamma \upharpoonright x + 1} \wedge (y) (y \leqslant x \supset \mathbf{CON}_{\gamma \upharpoonright y}))$. Thus from the definition of '$\gamma_1$' it follows that $(\exists x) ((y)(\gamma_1(y) \equiv \gamma(y) \wedge y \leqslant x) \wedge \mathbf{CON}_{\gamma \upharpoonright x})$. Thus $\mathbf{CON}_{\gamma 1}$. Hence in either case $\mathbf{CON}_{\gamma 1}$. It should be pointed out that the assumption that $\Gamma$ is consistent is used tacitly in the proof that '$\gamma_1$' bi-numerates $AX_\Gamma$. This is because the definiens of the definition of 'numerate' is a bi-conditional: if $\Gamma$ were inconsistent then $\vdash_\Gamma \gamma_1 (\tilde{n})$ whether or not $AX_\Gamma$ (n), and '$\gamma_1$' would not numerate $AX_\Gamma$. Moreover, as part (b) of the proof shows, if $\Gamma$ is not consistent then '$\gamma_1$' bi-numerates the Gödel numbers of some proper (possibly empty) subset of the axioms of $\Gamma$ which are necessarily consistent. As a result anyone who wanted to use Theorem 2 to answer scepticism about the consistency of some $\Gamma$ would have to show that $\mathbf{CON}_{\gamma 1}$ 'expresses' the consistency of $\Gamma$. However, he would have no hope of doing this unless he could establish the consistency of $\Gamma$ independently of Theorem 2.[19]

## 4. AN OPEN PROBLEM

The present analysis establishes that Feferman's result does not have the philosophical implications that a first reading would suggest. However, other (possibly deeper) problems have now surfaced. The foremost of these is the matter of distinguishing 'proper' or 'natural' consistency sentences from 'pathological' ones such as Feferman's. Of course, you cannot just 'read' $\mathbf{CON}_\gamma$ and decide that it 'means' that $\Gamma$ is consistent. The connection between $\mathbf{CON}_\gamma$ and the meta-language is dependent upon the particular Gödel numbering used. Thus the 'meaning' of $\mathbf{CON}_\gamma$ is relative to a Gödel numbering. But, of course, $\mathbf{CON}_{\gamma 1}$, the pathological sentence, is infected with the same problem. Clearly the reason that $\mathbf{CON}_\gamma$ strikes us as natural while $\mathbf{CON}_{\gamma 1}$ appears odd, is that the inner structure of the former more closely reflects our meta-mathematical definition of consistency. Moreover, the two sentences differ only in their components '$\gamma$' and '$\gamma_1$', which numerate $AX_\Gamma$. Thus the problem of determining 'natural' consistency sentences consists in describing the proper substitution instances for $\varphi$ in $\mathbf{CON}_\varphi$. In other words, the problem reduces to determining 'natural' numerations of the axioms of $\Gamma$.

If we inspect known axiom systems, we find that the axiom predicates (under Gödelization) have the form

$$AX_\Gamma(x) \equiv [x = n_0 \lor x = n_2 \lor \ldots \lor x = n_m \lor \Gamma_1(x) \lor$$
$$\Gamma_2(x) \lor \ldots \Gamma_p(c)] \ (m \geq 0, p \geq 0)$$

where each $n_1$ is a Gödel number of an axiom and each $\Gamma_1$ is (the code for) a schematic description of an infinity of axioms. Feferman has shown that no 'pathology' arises as long as $\Gamma$ has finitely many axioms.[20] Thus the problem can be reduced to defining the concept of *axiom schemata*.

The usual approach is to say that an axiom schema is an effective device for generating an infinite class of formulas.[21] Using Gödelization one then usually appeals to Church's thesis and requires (only) that the class of Gödel numbers of axioms be recursive (or even only recursively enumerable). In other words the condition of recursiveness is imposed upon the extensions of $\Gamma_1$. But this condition is too liberal since both '$\gamma$' and '$\gamma_1$' are bi-numerations of recursive sets.

At this point someone might recall that a set is recursive if and only if *there is* a recursive characteristic function for it and that the italicized quantifier can be interpreted constructively. Thus pathology might be prevented by insisting that the recursiveness of $\Gamma_1$ be demonstrated constructively. The bearing of this observation upon our problem is unclear, as the next example shows.

Consider the two number theoretic predicates $AX_p$ and $AX'_p$ where $AX_p$ is the usual axiom predicate for P and $AX'_p$ is defined after Feferman by

$$AX'_p(x) . \equiv . AX_p(x) . (y)(y \leqslant x \supset CON_{p \restriction \gamma}),$$

where $P \restriction y(z) . \equiv . AX_p(z) . z \leqslant y$. Gödel presented a constructive proof of the recursiveness of $AX_p$ in the course of proving his first incompleteness theorem. The other predicate contains unbounded quantification (in the definition of $CON_{p \restriction \gamma}$) so the usual results on the construction of recursive predicates do not apply to it. Nonetheless we can prove that $AX'_p$ is recursive because we can show that it is co-extensive with $AX_p$. This follows from the consistency of P. Since there are constructive consistency proofs for P, one seems justified in concluding that we have a constructive proof that $AX'_p$ is recursive. If this conclusion is correct then the requirement of constructive recursiveness will not serve to exclude pathological axiom predicates.

Still there is something odd about the 'proof'. Let $AX_{ZF}$ be the usual recursive axiom predicate for ZF-set theory. Let $AX'_{ZF}$ be defined analogously to $AX'_p$. We can prove that $AX'_{ZF}$ is recursive as follows: either ZF is consistent, in which case

$AX_{ZF}$ and $AX'_{ZF}$ are co-extensive. Or ZF is not consistent in which case $AX'_{ZF}$ has a *finite* extension. In either case the extension of $AX'_{ZF}$ is recursive. This proof is not constructive, however, since it assumes that ZF is or is not consistent. Moreover, we cannot apply our former proof technique to $AX'_{ZF}$ since it is not possible to give a constructive consistency proof for ZF — even in the broadest sense of 'constructive'.

The 'proof' of the recursiveness of $AX'_p$ is very atypical of proofs of recursiveness, and it does not generalize to other $AX'_\Gamma$ having the same logical and arithmetical form. This is perplexing. Clearly further investigation of the problem of constructivity is needed before these matters can be sorted out.

Perhaps the conditions on the axiom predicates must be given in terms of the syntactic form or their intensions. The latter alternative is repellent not only because it is vague but also because it introduces intensionality into mathematics. The obvious move, then, is to require that predicates be such as to produce RE-numerations upon formalization. However, even the restriction to PR-numerations is too broad; for Feferman has demonstrated:

THEOREM 3: Suppose that $\Gamma_1$ and $\Gamma_2$ are two axiom systems, that $AX_{\Gamma2}$ and $AX_{\Gamma1}$ are recursively enumerable, and that $\Gamma_2$ is consistent. Then with each PR-predicate $\gamma_2$ which bi-numerates $AX_{\Gamma2}$ in P we can effectively associate a PR-bi-numeration, $\gamma_1$, of $AX_{\Gamma1}$ in P for which

$$\vdash_p \mathbf{CON}_{\gamma 1} \supset \mathbf{CON}_{\gamma 2}.^{22}$$

Hence even among the PR-predicates there lurk senses of consistency in which the relative consistency of two very disparate systems can be shown using only the methods of elementary number theory. As in the case of the previous pathological predicates, an examination of the definition of '$\gamma_1$' reveals that the consistency of $\Gamma_2$ is needed to show that '$\gamma_1$' does indeed bi-numerate $AX_{\Gamma1}$.

What we want are axiom predicates $AX_\Gamma$ whose extension is the same whether or not $\Gamma$ or some other system is consistent. For the moment we seem forced to take the following rather restrictive approach. We will admit only those axiom schemata which read "all wffs of the form ... are axioms" where the dots are filled by devices like '$A\supset(B\supset A)$' or '$(x)Fx\supset Ft$'. (This excludes such statements as "all tautologies are axioms".) For simplicity, let us assume that the only letters in these devices are schematic letters for wffs. Then we could say that (under Gödelization and formalization) an axiom schema is a predicate of a p.r. extension of P of the form

$$(\exists x_1) \dots (\exists x_n) (x_1 < a \wedge \dots \wedge x_n < a \; \mathbf{FML}_s(x_1) \wedge \dots \wedge$$
$$\mathbf{FML}_s(x_n) \wedge a = f(x_1, \dots, x_n)),$$

where $f$ is a primitive recursive function whose values, for Gödel numbers of wffs as arguments, are Gödel numbers of axioms.[23]

There remains the problem of justifying this or any other proposal. The situation is quite parallel to that of Church's thesis. In both cases we are trying to give a formal explication of a pre-formal concept. In neither case is a mathematical proof of the correctness of the explication possible, although both the pre-formal concept and its *explicans* belong to mathematics. We can only appeal to supporting evidence, and in the present case there is little of this available.

Another important parallel has been emphasized by Judson Webb. He states that the 'significance' of Church's *theorem* derives from Church's thesis. Analogously, the 'insignificance' of Feferman's Theorem 2 will derive from a thesis concerning the proper manner for formulating consistency statements.[24]

To put the problem in perspective it should be emphasized that 'natural' consistency sentences are easy to find for the systems which we have actually specified so far. One simply applies the well-known Gödel techniques and obtains sentences which Theorems 2 and 3 do not cover. The problem arises when we attempt to formulate general results about consistency proofs for arbitrary formal systems. This problem is serious both mathematically — fallacious proofs and false results have resulted from failures to take it into account[25] — and philosophically — much of the discussion below will remain vague pending a satisfactory solution.

## 5. CONCLUSION

We have seen that despite Feferman's results Gödel's second

theorem vitiates the use of Hilbert-type epistemological programmes and consistency proofs as a response to mathematical scepticism. Thus consistency proofs fail to have the philosophical significance often attributed to them.

This does not mean that consistency proofs are of no interest to philosophers. We know that a 'non-pathological' consistency proof for a system S will use methods which are not available in S. When S is as strong a system as we are willing to entertain seriously then a consistency proof for it will yield no epistemological gain. But in other cases philosophers might argue that the proof uses methods which are merely different from rather than stronger than those available in the system in question. This claim has been made, for example, in the case of the constructive consistency proofs for elementary number theory. Similar philosophical investigations can be made on relative consistency proofs, since these differ from each other in the principles they employ. For example, most relative consistency proofs can be carried out within elementary number theory, but without using the theory of real numbers, no one has been able to prove the consistency of Quine's ML relative to that of his NF.

What about the consistency of all mathematics or of some strong system for set theory? How do we answer the sceptic? Since here a convincing proof is not possible, we have established that the sceptic demands too much. We cannot be certain that our axioms are free from contradiction and must treat them as hypotheses which may be abandoned or modified in the face of further mathematical experience. This attitude is taken by many foundational workers who also go on to voice opinions about the *likelihood* that various systems are consistent. Since these opinions are variously supported by appeals to the clarity of the mathematical concept formalized, the existence or non-existence of 'weird' models for the system and actual empirical experience with the system, this is surely a fruitful area for philosophical research.

## NOTES AND REFERENCES

*I would like to thank William J. Thomas and Paul Ziff for their comments on an earlier draft of this paper.

1. Cf. Hilbert, 'Mathematische Probleme' (1900). Reprinted in Hilbert, *Gesammette Abhandlungen*, Springer, 1935. Also see a letter

to Frege translated in Frege, *On the Foundations of Geometry and Formal Theories of Arithmetic* (transl. by Kluge), Yale, 1971.

2. Cf. Resnik, 'The Frege-Hilbert Controversy' forthcoming in *Philosophy and Phenomenological Research*.

3. Frege, *Foundations*, 20.

4. Hilbert's finitism is expounded in his 'On the Infinite' translated in *From Frege to Gödel* (ed. by van Heijenoort), Harvard, 1967.

5. The symbols 'FML$_\Gamma$', 'PF$_\Gamma$' and 'neg' are meta-linguistic abbreviations of 'formula of $\Gamma$', 'proof in $\Gamma$ of' and 'negation of', respectively.

6. Cf. Mendelson, *Introduction to Mathematical Logic*, D. van Nostrand Co., p. 147.

7. This language consists of numerals for the natural numbers, symbols for recursive functions and logical symbols. It is important to distinguish between this (extended) portion of natural language and a formal system for number theory.

8. I have switched to lower case Greek letters to emphasize the difference between intuitive number theoretic predicates and their representations and also to conform to Feferman's notation used below.

9. Cf. Mendelson, *Introduction to Mathematical Logic*, pp. 131-5.

10. The doubt cast is this: the sentence $\Phi$a is viewed by Hilbert as a device for asserting infinitely many real sentences and a finitistic proof of it is a schema for proving each of its real instances. Thus one should think that if all real instances of $\Phi$a are finitistically provable then there should be a finitistic proof of $\Phi$a too. But in the case in question this is possible only through a proof of the consistency of $\Gamma$. (Another way of looking at this: the phenomenon of $\omega$-incompleteness casts doubt on Hilbert's view of schemata.)

11. There is one disadvantage which Hilbert's Programme does not have. Gödel's first theorem applies to the new 'real' sentences; so we cannot completely formalize the set of 'real' truths.

12. The situation is more complex if we admit non-constructive rules of inference. For example, suppose we added to one of the usual systems for number theory the rule: if A is true then $\vdash$ A. Then all true number theoretic statements expressible in the system would be theorems, and thus so would be the 'consistency sentences' for consistent systems.

13. Feferman's PR- and RE-predicates are 'translations' of these predicates into P. This is glossed here to simplify exposition. See Feferman, 'Arithmetization of Meta-Mathematics in a General Setting', *Fundamenta Mathematicae* LXIX (1960), 53.

14. This, of course, has a 'translation' in P. Cf. note 13.

15. Cf. Feferman, 'Arithmetization', p. 84, for references.

16. *Ibid.*, p. 66. Robinson's system is obtained from P by replacing the induction schema by the single axiom: $(\forall a)\, (a \neq 0 \supset (\exists b)\, (a = b'))$. It is introduced into the theorem to obtain greater generality.

17. *Ibid.*, p. 68.

18. Talk of co-extensiveness, is, strictly speaking, improper here, since the predicates belong to uninterpreted systems. However, the predicates are co-extensive in all models of S (if there are any). Judson

Webb in an excellent paper appears to think that '$(\forall a) [\gamma(a) \equiv \gamma_1(a)]$' is provable, but this is so only if $\mathbf{CON}_\gamma$ is also provable. Cf. Webb, 'Metamathematics and the philosophy of mind', *Philosophy of Science* 35 (1968), 77.

19. Webb ('Metamathematics', p. 177) and Thomas (in conversation) have remarked that it is difficult to derive any assurance from $\vdash_\Gamma \mathbf{CON}_\gamma$ since if $\Gamma$ is inconsistent all its sentences are theorems. But we have been discussing $\vdash_p \mathbf{CON}_{\gamma 1}$ — and at first sight this is exciting. Of course, often $\mathbf{CON}_\gamma$ can be proved in systems stronger than $\Gamma$.

20. Feferman, 'Arithmetization', pp. 59, 81. These pages show that in this case a 'natural' consistency statement can be found.

21. This is a broad definition but I want to allow such 'schemata' as 'every tautology is an axiom'.

22. Feferman, 'Arithmetization', p. 84. Another suggestion is to order the RE-$\gamma$ as follows:

$$\gamma \leqslant_p \gamma' \text{ if and only if } \vdash_p \mathbf{CON}_{\gamma'} \supset \mathbf{CON}_\gamma$$
$$\gamma <_p \gamma' \text{ if and only if } \gamma \leqslant_p \gamma' \text{ and not } \gamma' \leqslant_p \gamma$$

However, Feferman (p. 81) shows that if $\Gamma$ has infinitely many axioms, then there is no minimum in this ordering.

23. Cf. Mendelson, *Introduction fo Mathematical Logic*, p. 139.

24. Webb, 'Metamathematics', p. 177. Webb considers the thesis that $\mathbf{CON}\gamma$ is correct if and only if $\gamma$ is an RE-predicate, but we have seen that this will not work.

25. Cf. Feferman, 'Arithmetization', p. 84.

# VII

# On Interpreting Gödel's Second Theorem

Michael Detlefsen*

## 1.

In this paper I critically evaluate the most widespread philosophical interpretations of Gödel's Second Incompleteness Theorem. My approach is to say what I think is wrong with these interpretations as they presently stand, and, where possible, to try to indicate what would have to be achieved were those interpretations to be revived, though revival is not, in my opinion, a reasonable hope.

Sections 2–7 discuss that cluster of interpretations that I choose to call the sceptical interpretations of Gödel's Second Theorem (hereafter G2). In Section 8 I consider that interpretation of G2 which attributes its significance to some alleged ill effects it has on Hilbert's Programme. I shall argue there that G2 does not imply the failure of Hilbert's Programme.

## 2. SCEPTICAL INTERPRETATIONS OF G2

There are many who have taken the position that G2 somehow shows that any consistency proof for a theory $T$ of which it (G2) holds will have to make use of a premise-set that is more dubitable[1] than the premise-set of any proof constructible in $T$. Among those holding this position (or one similar to it) are E.W. Beth, Paul Cohen, A. Grzegorczyk, E. Nagel and J.R. Newman, and, most recently, M.D. Resnik. To illustrate this point I will cite passages from the writings of Beth, Cohen and Resnik, for in their writing we find unusually concise statements of the interpretation in question.[2]

In Beth and Cohen (respectively) we find the following assertions.

> ... according to this theorem [G2], the arguments needed in a consistency proof for a deductive theory are always in some respect less elementary than those admitted in the theory itself ...[3]

> ... [G2] implies that the consistency of a mathematical system cannot be proved except by methods more powerful than those of the system itself ...[4,5]

If taken literally, the claims of both Beth and Cohen are false. For surely G2 shows nothing about every 'deductive theory' or 'mathematical system', but, at most, shows something about those theories or systems to which it applies.[6] But if we allow Beth and Cohen this obvious restriction, their position is not patently false.

Neither, it may be thought, is their position patently epistemological. Why, it might be asked, should we take their remarks to be of an epistemological rather than a purely logical character?

My response is that it is more charitable to Beth and Cohen to take them as attempting to make an epistemological point than to take them as attempting to make a purely logical point. For if the phrases 'less elementary than' and 'more powerful than' are reasonably interpreted as expressing a logical relationship, then the claims of Beth and Cohen are palpably false, whereas this is not so if the phrases in question are taken to express an epistemic relationship.

On a reasonable logical reading, the remarks of Beth and Cohen would amount to saying that G2 shows that the premise-set of any consistency proof for $T$ (where $T$ is a theory to which G2 applies) is *logically* (deductively) more powerful than $T$ itself, where $T_1$ is logically more powerful than $T_2$ if and only if $T_1$ is an extension[7] of $T_2$ but $T_2$ is not an extension of $T_1$. This reading would have Beth and Cohen asserting the obviously false claim that G2 shows that the premise set $P$ of any consistency proof for $T$ is an extension of $T$, but $T$ is not an extension of $P$.

My suggestion is, then, that we interpret Beth and Cohen as

making an epistemological claim equivalent to that given in the opening paragraph of this section.

M.D. Resnik has recently proposed a view of G2 which bears certain affinities to the Beth–Cohen view. Yet despite the similarities, there is also an important difference between Resnik's position and that of Beth and Cohen. This difference consists in the fact that Resnik restricts the supposed epistemological impact of G2 to a smaller class of theories than it would appear to be restricted to by Beth and Cohen. The restriction is explicit in the following passage from Resnik.

We know that a 'non-pathological' consistency proof for a system $S$ will use methods which are not available in $S$. When $S$ is as strong a system as we are willing to entertain seriously, then a consistency proof for it will yield no epistemological gain.[8]

Because of this notable difference between Beth–Cohen and Resnik, the views will be given different criticisms.


## 3. IS THE BETH–COHEN INTERPRETATION EPISTEMOLOGICALLY INTERESTING?

If the Beth–Cohen interpretation is to serve as a basis for attributing epistemological importance to G2, it would appear to stand in need of supplementation, for taken by itself, it would seem to be of little or no epistemological interest. We must try, then, to uncover certain additional premises needed to span the distance between the Beth–Cohen view and some interesting epistemological conclusion.

The conclusion toward which writers like Beth and Cohen[9] seem to be pressing is, as was mentioned in the opening section, what might properly be called a sceptical conclusion. For according to this conclusion there is an important restriction on how belief in the consistency of a theory might be justified; that is, such belief cannot be justified via proof of the usual formalizable variety. And, judging by our present lights, that amounts to saying that belief in the consistency of $T$ cannot be justified by proof — period. It remains for us to try to fill the gap between the Beth–Cohen interpretation and this explicitly interesting conclusion in as plausible a way as seems possible.

My suggestion is that, for want of an equally satisfactory alternative, we take the following pair of premises as furnishing the requisite supplementation.

SUP 1: If the premise-set of every consistency proof for $T$ is more dubitable than the premise-set of any proof in $T$, then the premise-set of every consistency proof for $T$ is more dubitable than the axiom-set for $T$.

SUP 2: If the premise-set of every consistency proof for $T$ is more dubitable than the axiom-set of $T$, then the premise-set of every consistency proof for $T$ is incapable of removing rational doubt or indecision from its conclusion.

If one supplements the Beth–Cohen view with the above pair of theses one gets to what looks like an interesting epistemological result. Without them, the epistemological importance of that view is not clear.

SUP 1 occupies a unique place in my criticism of Beth–Cohen. For either SUP 1 is an acceptable claim or it is not. If it is, then, as I shall shortly argue, it can be used to generate an attack on Beth–Cohen. If it is not, then Beth–Cohen is not worth attacking, since without the help of SUP 1, it is of little or no philosophical interest.

One could, I think, give quite a compelling critique of Beth–Cohen just by developing the above-mentioned dilemma. But my critique shall go further. I shall indeed argue for SUP 1. This I do with the thought that should the defender of Beth–Cohen fill the gap between it and some epistemologically interesting result (by means currently unforeseen) without appealing to SUP 1 (or some claim which entails it), then I would still have an argument against Beth–Cohen to fall back on.

## 4. AN ARGUMENT FOR SUP 1

The reader may already have observed that SUP 1 is trivially true for the class of finitely axiomatizable theories. If $T$ is finitely axiomatizable, then the entire axiom-set of $T$ will be the premise-set of various proofs in $T$. Thus, if the premise-set of any consistency proof for $T$ is more dubitable than the premise-set of any proof in $T$, then it is, by that very fact, more dubitable than the axiom-set of $T$.

However, it is absolutely crucial to notice that the reasoning which has just been used to show the plausibility of SUP 1 for finitely axiomatizable theories cannot be used to show its plausibility in the case of non-finitely axiomatizable theories. The reason for this is plain: in non-finitely axiomatizable theories, the entire axiom-set is never the premise-set of a proof in the theory. Because of this, a non-finitely axiomatizable theory will have to be shown to possess a very special sort of epistemic organization if one is to be able to show that SUP 1 holds for it.

There is a special type of epistemic organization which $T$ may possess and which enables us to demonstrate SUP 1 for $T$. This type of organization will be called 'epistemic compactness'.

We shall say that $T$ is 'epistemically compact' when the dubitability of the whole theory is, in a sense, 'reflected' in a finite portion of the theory. More precisely, we shall say that $T$ is epistemically compact when there is some finite subset $T_f$ of the axioms of $T$ that is as dubitable as the entire axiom-set of $T$.

The basis for my positive defence of the compactness thesis is the simple observation that often all one needs to know, and, in certain instances, all one can know, about a sentence $A$ is information regarding its 'form' (e.g., its logical form, arithmetic form, set-theoretic form, etc.). Thus, in certain cases my only justification for believing a sentence $A$ will be my belief that it has the (logical) form '$Bv-B$'. This will probably be true for cases where $B$ is very long or complicated and for cases where $B$ is some sentence undecided (and perhaps undecidable) by current knowledge. Similarly, I often commit myself to sentences whose content I've never examined and never will examine simply because they have a certain 'form'; for example, the 'form' prescribed by the axiom-schema of induction in $Z$, or the 'form' prescribed by the Aussonderungs principle of $ZF$. Indeed, tacit in the practice of using axiom-schemata to specify a theory is the assumption that the 'formal' information conveyed by the schema is both (i) all that we *need* to warrant acceptance of an instance of the schema and (ii) at least for certain instances, all the justification that we shall actually have for accepting those instances.

Now if we let '$In(AS)$' stand for an instance of axiom-schema $AS$ of $T$ that is justified and whose sole justification is the belief that it is of the 'form' prescribed by $AS$, then we get a

135

reason for believing in the epistemic compactness of $T$. For under those circumstances, our confidence in '$In(AS)$' is only as high as our confidence in the claim that all instances of $AS$ are true.[10] Thus, our confidence in $\{\{T_n\} \cup In(AS)\}$ will be no greater than our confidence in $T$ entire. ($\{T_n\}$ is the set of axioms of $T$ not given by $AS$.)

The preceding argument gives us the means to defend SUP 1 for non-finitely axiomatizable theories that are compact.[11] For if $T$ is given by the individual axioms $A_1, \ldots, A_n$ and the axiom-schemata $AS_1, \ldots, AS_k$, then the finite set $T_f$ of axioms of $T$ comprised of $A_1, \ldots, A_n$ and $In(AS_1), \ldots, In(AS_k)$ (where each $In(AS_i)$ is an instance of $AS_i$ whose sole justification is the general claim that all instances of $AS_i$ are true) is as dubitable as the axiom-set of $T$. Furthermore, $T_f$ is the premise-set of some proof in $T$. So if the premise-set of every consistency proof for $T$ is more dubitable than the premise-set of every proof in $T$, then the premise-set of every consistency proof for $T$ is more dubitable than $T_f$. And if the premise-set of every consistency proof for $T$ is more dubitable than $T_f$, then the premise-set of every consistency proof for $T$ is more dubitable than the axiom-set for $T$. Thus, SUP 1 is true for finitely axiomatizable theories and also for non-finitely axiomatizable theories that are epistemically compact.

Now there is another type of epistemic organization which a non-finitely axiomatizable $T$ may possess which does not entail $T$'s compactness. I shall call this sort of organization 'epistemic paracompactness'.

Think of a non-finitely axiomatizable theory $T$ which has a set $\{T_n\}$ of individual axioms plus an axiom-schema $AS$. And let $\{\{T_n\} \cup \{AS\}_j\}$ stand for the theory obtained by adding all of the instances up to and including the $j + 1$th ($j \geq 0$) instance of $AS$ to $\{T_n\}$. We may then define the sequence $S$ to be the sequence where $S_0$ is the dubitability value of $\{T_n\}$ and where $S_{m+1}$ is the dubitability value of $\{\{T_n\} \cup \{AS\}_m\}$. Then we shall say that $T$ is epistemically paracompact if the sequence $S$ converges at the dubitability value of $T$.

Now it would seem that $S$ is a monotone increasing sequence since, surely, as we continue to pile instances of $AS$ onto $\{T_n\}$ we keep getting sets of axioms of $T$ that are at least as dubitable as the ones which preceded it in the 'piling on' process. Furthermore, $S$ would seem to be bounded by the dubitability value of the entire axiom-set of $T$. And indeed the dubitability value of

$T$'s entire axiom-set might seem to form a least upper bound on $S$. But if this is true, then $S$ converges at the dubitability value of $T$. And this being so, $T$ is epistemically paracompact.

There are two points concerning paracompactness which I should now like to note. One is that the paracompactness of $T$ does not sponsor a proof of SUP 1 for $T$, since for paracompact $T$, a set $C$ may be more dubitable than any finite subset of axioms of $T$ and still *not* be more dubitable than the entire axiom-set of $T$. Indeed, where $T$ is paracompact but not compact, the entire axiom-set of $T$ is itself such a $C$.

Secondly I should like to note that the paracompactness of $T$ will turn out to be a strong enough condition on $T$ to permit us to refute the Beth–Cohen thesis. This point will be developed in the next section.

Finally, before going on to a refutation of Beth–Cohen I should like to remind the reader that my overall strategy in this paper is to construct a dilemma for Beth–Cohen. I think that if Beth–Cohen is to be an interesting thesis, then one needs SUP 1. And if one needs SUP 1, then one needs epistemic compactness. If one finds insufficient reason to believe in compactness, then one should, to the extent of that insufficiency, doubt the significance of Beth–Cohen. By my argument involving compactness in the next section, all I am doing is trying to convince the reader that insofar as one has reason to believe in the compactness of $T$, one also has reason to reject Beth–Cohen as false. Of course, the reader may think that there is not good reason to believe in the compactness of $T$, and to that extent he may doubt my refutation of Beth–Cohen which appeals to compactness. But, since compactness would seem to be needed for Beth–Cohen to be significant, to the extent that the reader doubts my use of compactness to refute Beth–Cohen he should also doubt the significance of Beth–Cohen.


## 5. A CRITIQUE OF THE BETH–COHEN INTERPRETATION

My criticism of the Beth–Cohen interpretation will begin with a frontal attack, i.e., with an argument to the effect that it is literally false for a very wide range of mathematical theories. After finishing this phase of my criticism, I shall discuss various problems which will arise for one who might attempt to amend Beth–Cohen in such a way as to avoid the frontal attack.

But the frontal attack is of considerable importance because, if successful, it shows that the mere fact that $C$ is a set of sentences which *logically implies* Con($T$) cannot by any means be taken as evidence for the claim that $C$ is more dubitable than the premise-set of any proof constructible in $T$. If there is a reason to believe that a given consistency proof for $T$ will employ a premise-set $P$ that is more dubitable than the premise-set of any proof constructible in $T$, that reason cannot consist in that mere fact that $P$ logically implies Con($T$).

The first argument that I would like to present is what (for reasons that will become apparent) I call the 'Reflexivity Argument'. This argument uses the reflexivity of $Z$, $RA$ and $ZF$[12] and epistemic compactness to establish the falsity of Beth–Cohen for these theories. The argument is as follows.

(1) $Z(RA, ZF)$ is reflexive.

(2) By epistemic compactness, there is some finitely axiomatizable subtheory $Z_f(RA_f, ZF_f)$ of $Z(RA, ZF)$ such that the axiom-set of $Z_f(RA_f, ZF_f)$ is as dubitable as the axiom-set of $Z(RA, ZF)$.

(3) By (2) (and the way $Z_f(RA_f, ZF_f)$ are constructed),[13] Con($Z_f$) (Con($RA_f$), Con($ZF_f$)) is as dubitable as Con($Z$) (Con($RA$), Con($ZF$)).

(4) By (1), $\vdash_Z$ Con($Z_f$) ($\vdash_{RA}$ Con($RA_f$), $\vdash_{ZF}$ Con($ZF_f$)).

(5) By (4), there is some finite set $\Delta_z(\Delta_{ra}, \Delta_{zf})$ of axioms of $Z(RA, ZF)$ such that $\Delta_z \vdash$ Con($Z_f$) ($\Delta_{ra} \vdash$ Con($RA_f$), $\Delta_{zf} \vdash$ Con($ZF_f$)).

(6) By (5), Con($Z_f$) (Con($RA_f$), Con($ZF_f$)) is not more dubitable than $\Delta_z(\Delta_{ra}, \Delta_{zf})$.

(7) By (3), (6), Con($Z$) (Con($RA$), Con($ZF$)) is not more dubitable than $\Delta_z(\Delta_{ra}, \Delta_{zf})$.

(8) By (7), Con($Z$) (Con($RA$), Con($ZF$)) is not more dubitable than the premise-set of any proof in $Z(RA, ZF)$.

(9) By (8), Beth–Cohen is false for $Z$, $RA$, $ZF$.

One can also show the Beth–Cohen view to be literally false for a host of theories 'weaker' than $Z$. This class of theories is what I shall refer to as the finite $Q$ extensions.[14] The argument is as follows.

(1) By the compactness of $Z$, there is some finite $Q$ extension $Q^*$ such that the axiom-set for $Q^*$ is as dubitable as that for any finite $Q$ extension.

(2) By the reflexivity of $Z$, the axiom-set of some finite $Q$ extension $Q^{**}$ (not identical to $Q^*$) logically implies $Con(Q^*)$.

(3) By (1), the axiom-set of $Q^*$ is as dubitable as that of $Q^{**}$.

(4) By (2), (3), there is a proof of $Con(Q^*)$ whose premise-set is not more dubitable than the premise-set of every proof in $Q^*$ (i.e., Beth–Cohen is false for $Q^*$ and every finite $Q$ extension that is an extension of $Q^*$).

I would now like to sketch an argument against Beth–Cohen *not* employing the notion of epistemic compactness. The argument that I have in mind appeals to only two features of $T$, namely (a) the paracompactness of $T$ and the claim that (b) $Con(T)$ is at least slightly less dubitable than $T$. If (a) and (b) both hold, then there is some finite set of axioms $T_f$ of $T$ such that the dubitability of $T_f$ comes closer to that of $T$ than does the dubitability of $Con(T)$. That being so, one may take $T_f$ and obtain the premise-set of a proof in $T$ that is more dubitable than some set of sentences implying $Con(T)$. This, then, defeats Beth–Cohen even for theories that are *not* compact if they meet (a) and (b) above.

Now one may think that assuming $T$ to meet (b) implies the falsity of Beth–Cohen for $T$. And this being so, it may be felt that the argument just given begs the question. But this is wrong. For (b) only implies the falsity of Beth–Cohen when (a) is present. In other words, one can only get from the assumption (which is (b)) that $Con(T)$ is less dubitable than the entire axiom-set of $T$ to the claim (which is the denial of Beth–Cohen for $T$) that some set of sentences implying $Con(T)$ is *not* more dubitable than each premise-set of a proof in $T$, if one also has (a).

Two further points are worthy of notice. In the first place, it should be pointed out that, at least for a large class of philosophers of mathematics, namely, those who are platonists, (b) is an entirely reasonable assumption to make. For according to the platonist, there is more constraining truth in mathematics than mere consistency. Secondly, it should be noted that the argument from (a) and (b) is effective not only against a literal reading of Beth–Cohen, but also a reading of it which substitutes the phrase 'as dubitable as' for the somewhat stronger phrase 'more dubitable than'.

## 6. REVISING THE BETH–COHEN INTERPRETATION

A further look at the reflexivity argument suggests a strategy for revising the Beth–Cohen interpretation in such a way as to preserve its 'spirit' and at the same time free it from at least some of the difficulties discussed above. For the proof of $\text{Con}(T)$ with which the reflexivity argument directly counters Beth–Cohen is the 'one liner' proof of $\text{Con}(T)$. This proof is, of course, trivial in a certain epistemic sense, viz. that anyone having rational doubt or indecision concerning the truth of its conclusion will have as much rational doubt or indecision with respect to its premise-set.

Noting this, the advocate of the Beth–Cohen interpretation might attempt to restate his position along the following lines.

> BCR: G2 shows that any 'non-trivial' consistency proof for theory $T$ will have to make use of a premise-set that is more dubitable than or as dubitable as the premise-set of any proof constructible in $T$.

Although BCR escapes the grasp of the Reflexivity Argument, it meets with other difficulties. To begin with, the argument concerning finite $Q$ extensions given earlier, though not a clear-cut counter-example to BCR, is nonetheless a troublesome case. It raises a challenge to BCR, namely, to give reason for believing that the proofs of $\text{Con}(Q^i)$ (where $Q^i$ is either $Q^*$ or a finite $Q$ extension that is an extension of $Q^*$) are 'trivial'. If all such proofs are 'trivial', it is not in the least obvious that they are. As a result, until such a time as the advocate of BCR can show us that all of the proofs of $\text{Con}(Q^i)$ are trivial, BCR will remain groundless for those cases of $T$ comprised of $Q^*$ and various of its extensions.

Another, and related problem confronting the advocate of BCR is to come up with support for BCR even in the case of $Z$, $RA$, $ZF$ (and various extensions of each). For even in these cases BCR would appear to be groundless.

Grounds there would be if it could be shown that *every* set of sentences outside[15] of $T$ which implies $\text{Con}(T)$ is either as open to doubt as or more open to doubt than every set of sentences inside of $T$. But that thesis would seem to be falsified by $\{Con(T)\}$ itself, as the Reflexivity Argument suggests.

And changing 'implies' to 'non-trivially implies' does not seem to help matters much. For there would seem to be no grounds of an *a priori* sort to suggest that just because a set of sentences lies on the exterior of $T$ and 'non-trivially' implies $Con(T)$ it will therefore be more dubitable than any finite set of sentences in the interior of $T$. Mathematical theories have not been consciously structured with such an end in mind.[16]

Nor would there appear to be any compelling empirical or inductive support for BCR. If we let $A$ be 'G2 holds for $T$', $B$ be 'all proofs of $Con(T)$ are more dubitable than any proof in $T$' and $C$ be 'all "non-trivial" proofs of $Con(T)$ are more dubitable than or as dubitable as any proof in $T$', then it seems correct to say that $C(A, C) \nleqslant C(A, B)$ (where $\ulcorner C(A, B) \urcorner$ is to be read 'the credibility of $B$ given $A$').[17] This is due to the fact there are such results as Gentzen's consistency proof for $Z$ and the Gödel–Gentzen proof of the consistency of $Z$ using $Con(H)$ and $H$ and also due to the compound fact that BCR is challenged by the finite $Q$ extension argument and that we see no reason to suppose that what happens with the finite $Q$ extension won't also happen with the other theories for which G2 holds. In general, the problem for the advocates of BCR attempting an empirical defense is to show us that $A$ is 'relevant' (epistemically) to $C$ in a way that it is not 'relevant' to $B$. For the reasons just discussed, I think this would be quite difficult, if not impossible, to do for any appreciable range of theories for which G2 holds.

## 7. RESNIK'S INTERPRETATION

In Resnik's interpretation of G2 (cf. Section 2 of this paper for a statement of this view) the troublesome notion seems to be that of 'a system as strong as we are willing to entertain seriously'. Resnik makes the claim that for such a system $S$, a consistency proof (of the 'non-pathological' variety) will yield no epistemological gain. Resnik's claim is, I think, interesting because of the fact that it seeks to restrict the class of theories for which G2 has epistemological significance to those theories which are 'as strong as we are willing to entertain seriously'. To

141

my knowledge, no comparable restriction is attempted by any other interpretation of G2.

That Resnik makes this studied restriction suggests that he thinks that there is something special about systems as strong as we are willing to entertain seriously that makes them, in a way not pertaining to other theories, epistemologically interesting targets for G2.

Let us call the theories to which Resnik restricts his claim $R$-theories. And let us see whether we can uncover any properties of $R$-theories that would serve to make them special.

One might want to characterize $R$-theories as those theories that are as dubitable as any theory which we are willing to entertain seriously. This is to give a decidedly epistemic reading to Resnik's phrase 'as strong as'. But so viewed, there would seem to be nothing separating $R$-theories from other theories in terms of the significance of G2. For, where $T$ is as dubitable a theory as we are willing to entertain seriously, it would seem no more plausible to believe that there is no statement $A$ outside of $T$ that serves as a 'non-trivial' proof of $Con(T)$, than it would to believe this where $T$ is not such a theory.

What might seem to be a more refined view of what an $R$-theory is, is given by the following definition. $T$ is an $R$-theory just in case (i) for every $T'$ different from $T$ if $T'$ is (deductively) an extension of $T$, then $T'$ is too dubitable to be entertained seriously and (ii) $T$ is entertained seriously. The question then is whether for $R$-theories so characterized, it is plausible to believe that consistency proofs will yield no epistemological gain.

It would not seem possible to sustain a positive response to this question via some sort of *a priori* defence; i.e., by some defence attempting to use the mere notion of an $R$-theory plus an appeal to G2 to generate the claim that consistency proofs will yield no gain. For in order to construct such a defence, one needs *a priori* assurance that every set of sentences $S$ outside of $T$ which logically implies $Con(T)$ does so 'trivially' (i.e., the initial dubitability of $S$ is as great as or greater than that of $Con(T)$). But surely one has no such assurances *a priori*.

In fact, there is considerable reason to doubt whether, so characterized, there are any $R$-theories. One (but not, I think, the only) reason for doubting the existence of $R$-theories is the following. $T \cup Con(T)$ is an extension (deductively) of $T$, but it seems clear that the dubitability of $T \cup Con(T)$ is no greater than that of $T$. And this means that if $T$ is a theory that we are willing to entertain seriously, then so also is $T \cup \text{Con}(T)$. If this is so, and it seems plausible enough, there could be no theory satisfying both clauses (i) and (ii) of the present definition of $R$-theories.

What Professor Resnik would seem to need is a theory $T$ so constructed as to make every set of sentences on its exterior which implies $\text{Con}(T)$ to be either (i) as dubitable as or more dubitable than any set of sentences in its interior or (ii) otherwise incapable of 'non-trivially' implying $\text{Con}(T)$. But such a theory presents some special problems for Resnik. First, in what sense could such a theory be said to be a theory 'as strong as we are willing to entertain seriously'? Secondly, what guarantee is there that such a theory would be of any importance to mathematics? Thirdly, and perhaps most importantly, what reason is there to think that such a theory would be recursively axiomatizable or otherwise capable of representing its proof theory and hence a theory for which G2 holds?[18]

Unless one can come up with the right answers to at least the latter two questions, it would seem that there is no hope of defensibly attributing epistemological significance to G2 on the grounds that there is some theory $T$ whose exterior is (i) at each 'point' either more dubitable than or as dubitable as each 'point' of its interior or (ii) otherwise incapable of 'non-trivially' implying $\text{Con}(T)$.[19]

And even if one could find the right answers to the latter two questions for a certain $T$, it is not clear that this would be of any aid to Resnik, for he would still be confronted with the task of showing for that $T$ that, in some plausible sense, it was 'as strong a system as we are willing to entertain seriously'. As I doubt that this could be done, I also doubt that Resnik's restriction (to $R$-theories) is at all relevant to the quest for a plausible and epistemologically significant interpretation of G2.

## 8. G2 AND HILBERT'S PROGRAMME

In this section I would like to sketch an argument against the claim that G2 implies the failure of Hilbert's Programme for finding a finitistic consistency proof for the various theories of classical mathematics.[20] The central claim of the argument is that $\text{Con}(T)$, the consistency formula shown to be unprovable by G2, does not really 'express' consistency in the sense of that term germane to an evaluation of Hilbert's Programme.

In order for a consistency formula to 'express' consistency in the appropriate sense the quantifiers and operators in it must be construed finitistically, and *not* classically, since it is the finitistic consistency of a classical system that is at issue. But a finitistic interpretation of the universal quantifier would seem to differ drastically from a classical interpretation of it, as is clear from the following remark of Herbrand.

> ... when we say that an argument (or theorem) is true for all (these) x, we mean that, for each x taken by itself, it is possible to repeat the general argument in question, which should be considered to be merely the prototype of these particular arguments.[21]

And, again, he says that a proof of a universal claim is merely a description or manual of the operations which are to be executed in each particular case.[22] This view of the universal quantifier would seem to sponsor the following restricted $w$-rule: if I have an effective procedure $P$ (i.e., a manual of operations $P$) for showing of each individual $n$ that '$F(\bar{n})$' is finitistically provable, then '$(x)F(x)$' is also finitistically provable. Indeed in a 1930 paper,[23] Hilbert stated a rule something like this. And at that time it was apparent to finitists that the rule did not give one the power to go beyond the means of some methods that had already been accepted as finitistic.[24]

Now one would not, in general, want to add the above-mentioned $w$-rule to a scheme designed to serve as the finitistic proof theory of the classical theory $T$, since that rule does not constitute a truth of the finitistic proof theory of the classical $T$! Still, certain instances of the rule would seem to be called for; in particular the one producing $\mathrm{Con}(T)$ from its instances. This addition made, $\mathrm{Con}(Z)$ becomes provable in $Z_{w^*}$ ($= Z$ plus the above-mentioned instance of the restricted $w$-rule).

Of course, if one adds instances of the restricted $w$-rule to $T$ in order to get an adequate context in which to do the finitistic proof theory of the classical $T$, then one will not be able to formulate the finitistic proof theory of $T$ as a formal system, but I see nothing in Hilbert's Programme which suggests that such formalizability is an essential or important feature of it. The essential thing is that $T$ itself be formalizable, since if this is not the case, the consistency of $T$ would not be a well-defined finitistic problem.

G2, then, only seems to imply the failure of Hilbert's Programme so long as one ignores the fact that the logic of the finitistic proof theory of the classical $T$ and the logic of the classical $T$ itself are two quite different logics! Once this is recognized, the fact that $\text{Con}(T)$ is not provable in $T$ should come as no particular shock to those espousing Hilbert's Programme. If the logic of $T$ is expanded in a way that produces a scheme whose logic is in agreement with the logic of the finitistic proof theory of the classical $T$, then in at least some instances (e.g., for the case where $T$ is the system $Z$), $\text{Con}(T)$ becomes provable. The basic flaw of those using G2 to thwart Hilbert's Programme is that they fail to recognize that the logic of the arithmetized proof theory of $T$ in G2 (since that arithmetized proof theory is itself embedded in $T$) is the logic of $T$ itself, *not* the logic of the finitistic proof theory of $T$ (which logic is *not* a subsystem of $T$'s logic)!

## 9. SUMMARY

In this paper I have considered various attempts to attribute significance to G2.[25] Two of these attempts (Beth–Cohen and the position maintaining that G2 shows the failure of Hilbert's Programme), I have argued, are literally false. Two others (BCR and Resnik's Interpretation), I have argued, are groundless.

## NOTE

*I would like to thank Dale Gottlieb, Stephen Barker, Tim McCarthy, Philip Kitcher, Michael Resnik and Richmond Thomason for extensive and helpful discussion of this work.

1. My use of dubitability is mainly an intuitive one. That is, I make use of only those aspects of the notion which seem to lie at the core of our intuitions concerning probability. I should, perhaps, also say that the notion of dubitability is taken here as applying to sentences and sets of sentences. When I speak of the dubitability of an interpreted sentence I intend to speak of the rational doubt and/or indecision concerning its truth. When I apply the notion to sets of interpreted sentences, I am intending to speak of the rational doubt and/or indecision concerning the joint truth of all sentences in the set.

I might also say that the notion of dubitability as I employ it could be

145

derived from the notion of epistemic probability as characterized, say, in Skyrms' *Choice and Chance*. So, our judgments of the truth of sentences in mathematics may change with time, and the probability of sentences of mathematics is *not* taken to be (at least in all cases) either 0 or 1 depending upon whether the sentence is false or true. There is nothing inherently irrational (rational) *per se* about mathematical falsehood (truth) in my view.

2. The appropriate sections in the writings of the other authors are Grzegorczyk, *An Outline of Mathematical Logic*, p. 576 and Nagel and Newman, *Gödel's Proof*, p. 6 for Grzegorczyk and Nagel–Newman respectively.

Actually Grzegorczyk takes a somewhat stronger and Nagel–Newman a somewhat weaker position than the one stated above. I will not, in the course of this paper, explicitly discuss either of these variations. Suffice it to say that everything I say about the present interpretations clearly applies to the stronger claim of Grzegorczyk. For a thorough discussion of the apparently weaker claim of Nagel and Newman, I refer the reader to Detlefsen's, 'The Importance of Gödel's Second Incompleteness Theorem for the Foundations of Mathematics'. pp. 71-78.

3. Beth, *Foundations of Mathematics*, p. 74, brackets mine.

4. Cohen, *Set Theory and the Continuum Hypothesis*, p. 3, brackets mine. Cohen makes a related remark in his 'Comments on the foundation of set theory', cf. p. 13.

5. In speaking of G2 as 'implying' such-and-such a conclusion $C$, Cohen can, I think, be taken as meaning that there is a set $P$ of plausible statements such that $P$ itself does not logically imply $C$, but $P \cup$ G2 does.

6. For our purposes it will suffice to think of G2 as applying to the consistent, recursively axiomatizable extensions of the well-known system $Q$. However, the reader should be aware of the fact that G2 actually holds for a considerably wider class of theories. For a characterization of this broader class of theories see Rosser, 'Gödel's theorems for non-constructive logics'. In the remainder of this paper when I speak of theories, I shall mean recursively axiomatizable theories, unless otherwise stated. I also take theories to be deductively closed sets of sentences.

7. $T_1$ is an extension of $T_2$, iff both $T_1$ and $T_2$ are deductively closed sets of sentences and $T_2 \subseteq T_1$. If theories are not regarded as closed then $T_1$ can be said to be an extension of $T_2$ when the closure of $T_2$ is a subset of the closure of $T_1$.

8. Resnik, 'The philosophical significance of consistency proofs', p. 145 and Chapter 6, this volume.

9. One finds a view like that of Beth–Cohen tempting Wang in his earlier writings, cf. Wang, *Logic, Computers and Sets*, p. 27. However, Wang seems to have turned his back on this view in his more recent writing, cf. Wang, *From Mathematics to Philosophy*, pp. 42-3.

10. An argument similar to this cannot be made when $In(AS)$ is not assumed to be justified. This is so because one circumstance that will lead to $In(AS)$'s not being justified is that the general claim is not

justified (assuming, of course, that $In(AS)$ has no 'individualized' justification; i.e. no justification that applies to $In(AS)$ but not to the other instances of $AS$). And as there can be good reason for doubting the truth of the general claim (i.e., the claim that all instances of $AS$ are true) that are not equally good reasons for doubting the truth of $In(AS)$, it cannot be expected that the dubitability of $In(AS)$ will match that of the general claim when $In(AS)$ is unjustified.

11. The reader will recall that 'theories' for me means, unless otherwise stated, 'recursively axiomatizable theories'.

12. A theory $T$ is said to be reflexive just in case for every finitely axiomatizable sub-theory $T_f$ of $T$, $\vdash_T \mathrm{Con}(T_f)$. $Z$ (first-order number theory), $RA$ (real arithmetic) and $ZF$ (Zermelo-Fraenkel set theory) were first proved to be reflexive by A. Mostowski in 'On models of mathematical systems'.

13. They are constructed by taking the finitely many individual axioms of $Z(RA, ZF)$ and adding to them for each axiom-schema in the theory, an instance of the schema whose sole justification is the claim that all of the instances of the schema are true, and then closing the set under deduction. That $\mathrm{Con}(Z_f)$ will be as dubitable as $\mathrm{Con}(Z)$ can be seen from the following argument.

Suppose that we divide the dubitability of a theory into two parts: (i) consistency worries and (ii) extra consistency worries, or 'factual' worries, as I shall call them. Now the factual worries concerning $Z_f$ cannot exceed those of $Z$ for every factual worry concerning $Z_f$ is ipso facto one for $Z$. But, since the dubitability of $Z_f$ equals that of $Z$ (by epistemic compactness), and the 'factual' worries concerning $Z_f$ cannot exceed those concerning $Z$, the consistency worries of $Z_f$ must be equal to those of $Z$. In short, the consistency worries concerning $Z_f$ cannot be greater than those for $Z$. And if they were less, the factual worries concerning $Z_f$ would be greater than those concerning $Z$, which they cannot be. Thus, the consistency worries concerning $Z_f$ equal those concerning $Z$. And what has been done for $Z$ can obviously be repeated for $RA$ and $ZF$. Thus we get (3).

14. A finite $Q$ extension will be any theory $T$ meeting the following conditions.

(i) the axioms of $Q$ are axioms of $T$

(ii) $T$ has some (but not more than a finite number of) instances of the induction schema as axioms

(iii) $T$ has only those axioms provided for by (i) and (ii).

15. Since theories for me are deductively closed sets of sentences, a sentence being on the exterior of $T$ (or outside of $T$) simply means that the sentence isn't a theorem of $T$. Similarly, but oppositely, for sentences in the interior of $T$.

16. Indeed proofs such as Gentzen's of the consistency of $Z$ which can be formalized in $H$ with a constructivistic $w$-rule and Gentzen's and Gödel's of the relative consistency of $Z$ to $H$ (intuitionistic number theory) would seem to give at least some support to the view there are proofs of $\mathrm{Con}(Z)$ on the exterior of $Z$ which non-trivially imply $\mathrm{Con}(Z)$ and which are not as open to rational doubt as every finite set of theorems of $Z$.

17. Credibility is just to be treated as a cognate of dubitability. The more credible a sentence, the less dubitable it is.

18. Rosser, 'Gödel theorems for non-constructive logics', showed that there are systems that can represent their proof theory even though these systems are not recursively axiomatizable using ordinary quantificational logic.

19. A 'point' on the exterior of $T$ is any set of sentences of the language of $T$ that is not a subset of $T$. A 'point' in the interior of $T$ is any set of sentences that is a subset of $T$.

20. See *Hilbert's Program* (Dordrecht, D. Reidel Publishing Co., 1987).

21. In Goldfarb, *Jacques Herbrand: Logical Writings*, pp. 288-9, fn. 5.

22. Ibid., pp. 49-51.

23. See Hilbert, 'Die Grundlegung der elementaren Zahlenlehre', reprinted in *Gesammelte Abhandlungen*, pp. 192-5.

24. In Goldfarb, *Jacques Herbrand: Logical Writings*, p. 297.

25. Conspicuously absent is any mention of how Feferman's work in his 1960 paper 'The Arithmetization of Metamathematics in a General Setting' might be called upon to reverse G2. I don't think it can, as I argued at the 1978 Western APA meeting in a paper entitled 'The Resolution of some Intentional Problems Concerning Gödel's Second Incompleteness Theorem'. That paper is currently being revised and expanded. A copy of the revised manuscript is available on request.

## BIBLIOGRAPHY

Beth, Evert W. (1968) *The Foundations of Mathematics*, 2nd ed. rev., North-Holland Publishing Co., Amsterdam.

Cohen, Paul J. (1966) *Set Theory and the Continuum Hypothesis*, W.A. Benjamin, Inc., Reading, Mass.

Cohen, Paul J. (1971) 'Comments on the Foundations of Set Theory'. In *Axiomatic Set Theory*, edited by Dana Scott, Proceedings of Symposia in Pure Mathematics, vol. 13, pt. 1, American Mathematical Society, New York.

Detlefsen, Michael (1976) 'The Importance of Gödel's Second Incompleteness Theorem for the Foundation of Mathematics'. Doctoral thesis, Johns Hopkins University.

Grzegorczyk, Andrzej (1974) *An Outline of Mathematical Logic*, D. Reidel Publishing Co., Boston.

Mostowski, Andrzej (1952) 'On Models of Axiomatic Systems'. *Fundamenta Mathematicae* 39, 133-56.

Nagel, E. and Newman, J.R. (1958) *Gödel's Proof*, New York University Press, New York.

Resnik, Michael D. (1974) 'The Philosophical Significance of Consistency Proofs'. *Journal of Philosophical Logic* 3, 133-47. Reprinted Chapter 6, this volume.

Rosser, J.B. (1937) 'Gödel's Theorems for Non-Constructive Logics',

*Journal of Symbolic Logic* 2, 129-37.

Wang, Hao (1970) *Logic, Computers, and Sets*, Chelsea Publishing Co., New York.

Wang, Hao (1974) *From Mathematics to Philosophy*, Humanities Press, New York.

# POSTSCRIPT

In a recent review (cf. *Mathematical Reviews* 80h: 03008) C. Smorynski has made some critical claims to which I should like to respond. He begins by stating that the goal of my paper is to refute what he calls the 'popular account' of G2. According to this popular account, G2 shows that a consistency proof for a sufficiently strong consistent theory must use methods more powerful than those of the theory itself. But, Smorynski tells us, this popular account admits of two different interpretations. One of these is a 'definite logical' reading in terms of the notion of relative interpretability, and presumably based on such facts as that T + Con(T) is not relatively interpretable in T. The other is an epistemological interpretation according to which G2 is taken to imply that a consistency proof for a theory will have to make use of a premise-set that is more open to rational doubt than the premise-set of any proof in the theory itself. Smorynski's charge is that I 'ignore' the former interpretation, and concentrate on the second interpretation, which he calls 'vague'.

Smorynski is correct in thinking that I intended to discuss a 'popular' account of G2. But the popular account I discuss is not the one he describes. For unlike his, mine is not ambiguous as between a logical reading in terms of relative interpretability and an epistemological reading like the one stated above. Rather, it is an unequivocally epistemological account, and is presented as such in the writings of Beth, Cohen, and various other writers cited in the paper. Perhaps this would have been clearer to the reviewer had I actually quoted some of the passages which were only cited in the paper (and if I had been able to present the full context of the remarks that actually were quoted). Consider, for example, the remark by Cohen (cited in footnote 4 of the paper) where he attributes the following effect to G2: 'There is always a great sense of frustration about consistency proofs and they seem to retain a circular character which makes them most unsatisfying' (Cohen, 'Comments on the Foundations of Set Theory', p. 13).

I find it hard to take seriously the idea that remarks such as this are really to be understood as making a purely logical claim about relative interpretability. Why, if relative interpretability is the real concern, is there never any mention of it? And why the use of such seemingly epistemological terminology as 'circular character'? The only 'logical' reading of these passages that has any textual support at all is the one I discuss briefly in section 2 of the paper; namely, that which takes them as saying that the premises needed for a consistency proof of a theory are actually deductively more powerful than the theory itself. But, as I mention, this is not a very charitable interpretation of the texts. And furthermore (modulo some idealizing assumptions) it implies the epistemological interpretation. Hence, it seems clear that the epistemological interpretation is the proper one to focus on.

The upshot of all this is that the 'popular' account of G2 that I claim to discuss in the paper (i.e., the account found in the writings of Beth, Cohen, and the others cited) really is an epistemological account, and not a logical account having to do with having overlooked the relative interpretability reading of G2. I intended to discuss the interpretation of G2 that is found in the texts referred to. The relative interpretability reading of G2 is not in those texts. Therefore, it isn't discussed. But that is only omission; not oversight. Moreover, given the philosophical character of my concern with G2, it seems an appropriate omission. For the relative interpretability reading of G2 is apparently without any philosophical interest whatever.

Let me now briefly consider the other two complaints that Smorynski makes. The first of these is that I 'conflate' the 'intensional and extensional notions of theories'. I have repeatedly pressed Smorynski to give me something like an argument for this claim, but he has so far failed to produce any. Consequently, I can only guess at what such an argument might look like.

In the paper I sometimes characterize theories (e.g., the finite Q-extensions) in terms of their axioms or (given the usual assumptions about their logic) theorems. This apparently led Smorynski to think that I *identify* a theory with a set of theorems (or a set of theorems as generated from a set of axioms via a set of derivations). And it would, of course, be wrong for me to do so, since both the reflexivity argument and the related argument concerning finite Q-extensions depend upon a view of theory which sees it not simply as a set of theorems or a set of theorems generated from a set of axioms via a certain set of derivations, but rather as a set of theorems *generated from* a set of axioms via a set of derivations *in a*

151

*particular way* (i.e., by means of a particular procedure or set of procedures for determining whether something is an axiom, derivation, or theorem). The character of these testing procedures for axiomhood, derivationhood, and theoremhood is sometimes an important consideration in determining whether certain metamathematical results hold for a given body of axioms, derivations, and theorems. And so it seems only appropriate that in connection with such results we should take the character of the testing procedures as one of the factors that determines the identity of a theory. Feferman (in his well-known paper 'The Arithmetization of Metamathematics in a Generating Setting') attached the term 'intensional' to such factors, and so a conception of theory which includes such factors among the identity conditions for theories is an 'intensional' conception of theory.

G2 is perhaps the best-known case of a result which depends upon an intensional view of theory. For as Feferman (and Rosser and Kreisel before him) pointed out, there are pairs of theories T and T′ having all and only the same axioms, derivations, and theorems where G2 holds for the one and fails for the other. Somewhat more precisely, there are pairs of theories T and T′ such that (1) the consistency formula for T is *not* provable in T, (2) the consistency formula for T′ (i.e., the formula constructed from the provability predicate for T′ in the same way that the consistency formula for T is constructed from the provability predicate for T) *is* provable in T′, and (3) T and T′ have exactly the same axioms, derivations, and theorems. Feferman's variant theories (cf., theorem 5.9 of the paper cited above) change the test procedure for axiomhood, while those of Rosser (as modified by Kreisel) change the test for derivationhood without changing that for axiomhood. But the basic message behind both types of variants is the same, namely, that some results of metamathematics utilize an intensional rather than an extensional notion of theory.

Like G2, Mostowski's proofs of the reflexivity of PA, R, and ZF utilize an intensional notion of theory since the provable consistency formula for the finitely axiomatized subtheories is assumed to be defined from the 'ordinary' provability predicate (which is taken to correspond to the 'ordinary' intensional notion of provability or derivability). Thus, since my reflexivity argument and the related argument concerning finite Q-extensions appeals to these results of Mostowski's, they themselves depend upon taking an intensional view of theories.

This may seem to be at odds with my practice (mentioned earlier) of characterizing theories in terms of their axioms or theorems. But given the prevailing conventions (specifically, that which leads one to assume that in intensional contexts it is the 'ordinary' notions of axiomhood, theoremhood, etc. that apply, unless otherwise stated), I don't think that it is. I was well aware of the importance of the distinction between intensional and extensional conceptions of theory when I wrote the paper. And I was also well aware that Mostowski's reflexivity theorems utilize an intensional conception. But I didn't feel any obligation to stress this fact because of the convention just mentioned.

I could cite many instances in the literature where other authors depend upon the same convention. But perhaps none is more appropriate than one by Smorynski himself. In his recent book *Self-Reference and Modal Logic* (Springer-Verlag) he says (p. 10) that G2 'tells' us that any true, sufficiently strong theory cannot prove its own consistency. Taken literally (that is without the beneficence of the convention mentioned above) this is false, and might be said to 'conflate' the extensional and intensional conceptions of theory. For the Rosser variant of PA, since it has exactly the same theorems as PA, it is both 'true' and 'sufficiently strong' (assuming that PA is). Yet G2 does not tell us that the Rosser variant of PA cannot prove its own consistency. Thus, in order for Smorynski's statement to be true, we must assume (in accordance with the common convention) that the theories of which he speaks employ the 'ordinary' intensional notion of provability. But if the same consideration is extended to my talk about theories in the reflexivity and finite Q-extension arguments, then I cannot be said to conflate the intensional and extensional notions of theory either. And so, I plead innocent to the charge of having done so.

As a final point, let me briefly respond to Smorynski's criticism of my attempt to sketch the idea for a defence of Hilbert's Programme against G2. The basic idea is a simple one; namely, that certain things that Hilbert (and other constructivists, such as Herbrand) said about universal generalizations suggest that finitary reasoning might allow for some sort of $\omega$-rule in its logic. If this were so, there would be a considerable problem for the traditional anti-Hilbertian argument from G2; namely, to show that the power of such an $\omega$-rule can be accommodated within the confines of a theory like PA whose

153

logic does not contain an $\omega$-rule. It is true that I did not give a precise statement of any $\omega$-rule and conclusively derive its justification from a finitary semantics for the universal quantifier. Rather, I only gave a rough indication of the sort of grounds upon which such a case might be made, and of how this might upset the traditional anti-Hilbertian application of G2. I aimed to be suggestive, and I think I was. Perhaps, however, I overstated my case. At any rate, I now believe that there are other, clearer, shortcomings in the traditional anti-Hilbertian argument from G2.

# VIII

## Wittgenstein's Remarks on the Significance of Gödel's Theorem[1]

S.G. Shanker

### 1. WITTGENSTEIN'S HERESY

Few attempts have been made to question the significance of Gödel's first incompleteness theorem. The fact that Wittgenstein appears to have pursued such a course has been a minor source of irritation to those mathematical logicians who look to his works for the 'sparkles of intelligence' they have found elsewhere in his writings; and a matter of acute embarrassment for Wittgenstein's critics who find themselves unable to explain what at best appears to be an uncharacteristically casual digression. Wittgenstein's argument suffers from several inescapable defects, not least of which is its opacity. It is clear that Wittgenstein was disturbed by what he saw as the philosophical consequences of Gödel's theorem; yet nowhere in his *Nachlass* is there a detailed critique of Gödel's proof to be found. As a result Wittgenstein's enigmatic objections in *Remarks on the Foundations of Mathematics* often appear completely unwarranted, even bizarre. To make matters worse, at one point Wittgenstein confessed that his task as he perceived it 'is not to talk about (e.g.) Gödel's proof, but to by-pass it' (RFM[2] VII §19). But how could any philosopher of mathematics seriously propose to bypass 'one of the most important advances in logic in modern times'?[3] Is it any wonder that such an apparently cavalier attitude towards one of the landmarks of mathematical logic should have met with such widespread condemnation? But perhaps these reactions are over-hasty, if not misconstrued. For whether or not Wittgenstein's criticisms hit their mark — forcing us, as Wittgenstein intended, to reassess our attitude to Gödel's theorem — is obviously a matter which we cannot hope to address until we have established the philosophical basis for Wittgenstein's arguments. And that is a task which takes us deep into the conceptual foundations of Gödel's thought.

Throughout the encomiums to Gödel's incompleteness theorem we find an unremitting attempt to convey some of the sublime quality of Gödel's proof. Not content with describing it as 'the most brilliant, most difficult, and most stunning sequence of reasoning in modern logic',[4] Nagel and Newman go a step further: it is nothing less than an 'amazing intellectual symphony'.[5] Perhaps, then, the philosophical appreciation of Gödel's theorem belongs as much to the philosophy of art as of mathematics? It is a question which should not be peremptorily dismissed; for mathematics may have one foot in the door of the sciences, but it bears asking where the other one rests. The 'Queen of the Sciences', like Bagehot's constitutional monarch, operates on a different level from the Parliament of Science. Hence the adjectives used to describe an important mathematical proof are fundamentally and necessarily different from those employed to characterize a successful scientific theory. It is certainly no idle whim that the term 'mystery' should so prefigure in discussions of scientific advances, whereas the incompleteness theorem should repeatedly be described as an 'intellectual symphony'. As vague as the latter metaphor might at first appear it is, in fact, one of the most revealing comments that could be made about the nature of Gödel's theorem, albeit for reasons other than are generally intended. For it points, not only to the profound aesthetic satisfaction which one experiences when working through Gödel's proof, but also to the very obstacles that must be confronted when trying to assess its overall significance.

Douglas Hofstadter, to mention but one example, has structured an imposing tome around this dimension of Gödel's proof, constantly striving to elucidate not just the mathematical but more importantly, the trans-mathematical significance of Gödel's theorem with analogies drawn from music and art.[6] But where Hofstadter's thesis encounters its greatest difficulty is precisely in the endeavour to describe the philosophical import of each. Does Bach's *Musical Offering* have any philosophical significance? Certainly it has a musical, but does it have a higher — perhaps a 'meta-musical' — meaning? So too, Gödel's theorem has unquestionable mathematical significance, but the question we must ask ourselves at the outset of any extra-mathematical evaluation is whether it is problematic to speak of the *philosophical* — or even the *meta-mathematical* — significance of Gödel's theorem.

Understandably, mathematicians have always been eager to review what they regard as their crowning achievements. But it can be a distressing sight to witness their suffering as they step outside the parameters of their proofs and venture into the hostile territories of philosophy. Such is the price that must be paid, however, for pursuing a field that hovers uncertainly between the sciences and the arts. What this tension illustrates is, perhaps, simply the limitations of prose for the elucidation of pure mathematical results. But herein also lies a matter of pressing importance to the vast number of mathematicians who populate our universities. For perhaps the greatest danger facing modern mathematicians is their sheer number: it has now reached the point where it is physically quite impossible to keep track of all the theorems published in any given week, let alone during the span of a year.[7] It is a disturbing thought that there might be monumental discoveries that have already been published, only to be consigned to the oblivion of the multitude of mathematical journals; or even worse, that will not even be given a reading by inundated editors. But if this seems a somewhat unrealistic danger, how much more worrying must it be for an aspiring young mathematician who is eager to attain some prominence in his career in these circumstances. Sadly for him, we are running out of Hilbert's Problems (and thus, vacancies in the so-called 'Honours Class' of mathematics). Clearly we are witnessing a massive institutional upheaval here as elsewhere; in the meantime, the prudent mathematical prodigy will choose his topics with great care. For he must try to settle on those problems where a success is most likely to gain notice. But how does one judge in advance how much distinction a proof might acquire?

It would help if, to begin with, he had a clearer idea of what contributes to the significance of an important mathematical theorem. To understand this he might begin by turning to Hilbert's 'Mathematical Problems', for this question is closely connected to the issue that Hilbert confronted when he pondered his Paris lecture in 1900. It is easy to see why Hilbert's paper fired the imagination of a generation of mathematicians; here, for the first time, was a serious inquiry into the 'general criteria which mark a good mathematical problem'.[8] The characteristics which Hilbert seized upon seem straightforward enough: a good problem should be clear, difficult but not inaccessible, and — in a curious relapse into platonist imagery — 'It should be to us a guide post on the mazy paths to hidden truths.'[9] Still, it might be questioned whether the influence of Hilbert's answer has been wholly salutary. For as a result of the fact that his attention was focused on responding to duBois-Reymond,[10] Hilbert ultimately created a distorted and in some ways damaging picture of mathematical significance. Hilbert's primary motive was to establish that there was no problem, however intractable it might at first appear, which could not be resolved if only the right point of view from which to attack it could be discovered. He thus chose his problems largely because they had resisted all attempts at a solution; not because of any intrinsic merit discernible in the problems themselves. And he did so for the very good reason that, as he himself remarked, 'It is difficult and often impossible to judge the value of a problem correctly in advance; for the final award depends upon the gain which science obtains from the problem.'[11] In other words, the *significance* of a mathematical theorem does not simply consist in providing the answer to a question which had hitherto eluded mathematicians. Or to put this another way, Hilbert's 'universal problem-status' may be a necessary (although even this is highly debatable) but it is not a sufficient condition for mathematical significance.

To see this we might consider a well-known example such as 'Do four 7s occur in the expansion of $\pi$?' Until fairly recently this was thought by many (philosophically minded) mathematicians to be an important unsolvable problem. The answer was, in fact, contained in a paper published by Daniel Shanks and John W. Wrench, Jr. in 1962, which simply consisted of a print-out of the expansion of $\pi$ up to 5,000 decimals.[12] It is surely no wonder that this achievement has been passed over in relative silence. As we shall see below, the significance of a theorem — no matter how long the problem in question might have remained unsolved — depends more than anything else on the manner in which the theorem is established; for it is this which determines 'the gain which science obtains from the problem'. But here too lurk complications similar to those which ensnared Hardy when he sought to articulate the value of pure mathematics.

A large part of *A Mathematician's Apology* is devoted to the vexing question of the usefulness of mathematics. The basic problem with Hardy's argument is that it proceeds from an orthodox utilitarian premise which Hardy labours throughout the text to overcome. He equated value with utility from the outset, but clearly the value of pure mathematics must lie in some other domain. For 'The "real" mathematics of "real" mathematicians, the mathematics of Fermat and Euler and Gauss and Abel and Riemann, is almost wholly "useless" (and this is as true of "applied" as of "pure" mathematics). It is not possible to justify the life of any genuine professional mathematician on the ground of the "utility" of his work.'[13] The ultimate solution which Hardy pursued was simply to repudiate the utilitarian premise which had inspired all of these difficulties in the first place: pure mathematics can only be 'justified as art if it can be justified at all'.[14] But then, what exactly does that mean: that mathematics (and art) can at best be justified in terms of some intangible spiritual benefit? Or is it rather a reflection of the fact that it is entirely misplaced to introduce the concept of *justification* here?

Obviously, we cannot speak of the mathematical significance of a theorem in the same way that a political theorist debates the implications of a social or ethical decision. Yet neither can we compare it to the discovery of a scientific law. In his attempt to come to terms with this problem Whitehead argued:

> The notion of the importance of pattern is as old as civilization. Every art is founded on the study of pattern. The cohesion of social systems depends on the maintenance of patterns of behaviour, and advances in civilization depend on the fortunate modification of such behaviour patterns. Thus the infusion of patterns into natural occurrences and the stability of such patterns, and the modification of such patterns is the necessary condition for the realization of the Good. Mathematics is the most powerful technique for the understanding of pattern, and for the analysis of the relation of patterns. ... Having regard to the immensity of its subject matter, mathematics, even modern mathematics is a science in its babyhood.[15]

This may be a subtle effort to circumvent the problem, but the solution we are searching for cannot be that there is no genuine difference between science and art and hence no problem in

159

placing mathematics in both camps. On logical grounds alone it is manifest that e.g. the propositions of fiction or the drawings of art are *toto caelo* different from the hypotheses of science or the theorems of topology. More to the point, perhaps, is the fact that the supposition that mathematics shares with the sciences the search for patterns in nature is a throwback to the metaphysical picture of the Laws of Nature which flourished during the Quattrocento. The real difference here is that where science searches for patterns, mathematics constructs them. And where the test for the significance of a scientific law is strictly empirical, there are no mathematical predictions or explanations whereby we can determine the import of a theorem.

The matter remains no less murky, however, when we turn to the realm of pure art for illumination. In an abstract domain such as music we can distinguish between two different types of pivotal work. If we ask about the significance of the *Prague* Symphony it is clear that the answer hinges on the manner in which Mozart carried classical Viennese style to a new plateau. For his intention was no longer purely pyrotechnic; rather, he created polyphonic unity from an extraordinary number of subjects, thereby rendering a work of unparalleled thematic richness in the genre. But the criteria we have in mind when describing a piece like Beethoven's *Eroica* as a revolutionary work (both in spirit and in design) are categorially different. Here we find an entirely new species of symphony: one that assaults the listener through the tempestuous emotions aroused, the dissonances, and the cycles of resolved tension which Beethoven created.

There are distinct echoes of the type of considerations operating here in the classification of significant proofs. Euler's theorem, for example, would be the *Prague* Symphony of function theory (in complexity if not in elegance), whereas Gauss' fundamental theorem would be the *Eroica* of algebra (marking as it does an entirely new attitude towards the nature of 'mathematical existence'). But according to Hardy, the mathematical significance of a theorem lies in the new ideas which it forges, and a mathematical idea is 'significant' if it can be 'connected, in a natural and illuminating way, with a large complex of other mathematical ideas'.[16] Ignoring for the moment any of the complications involved in Hardy's platonism it is at least evident from this that while there seems to be a striking parallel between the *Eroica* and Gauss' fundamental theorem — in terms of the revolutionary role played by each in the creation of a new school of musical/mathematical thought — it is nonetheless clear that the conceptual network that operates in the latter sharply distinguishes it from the former. Moreover, there is an element which renders one particular species of mathematics strikingly different from any aesthetic example, and far more akin to science. Indeed, on the type of approach advocated by Lakatos this feature is crucial to the classification of mathematics as a science: viz. the creation of impossibility proofs which result in the demise of a problem or an entire theory. And it is, of course, just this function which most sharply characterizes Gödel's 'intellectual symphony', rendering it that much more problematic to identify and elucidate the significance of Gödel's first incompleteness theorem.

In light of the phenomenon described above we should enter a note of caution here lest any highly impressionable tyro might be following this argument too closely: significance does not always guarantee success. One of the most momentous discoveries ever made in the history (or at least, apocrypha) of mathematics was Hippasus of Metapontum's construction of an incommensurable ratio: an achievement which, so legend goes, was met with summary execution. For Hippasus' crime was to prove that it is impossible to reduce every phenomena in the universe to whole numbers or their ratios; a matter of far more than mathematical importance as far as the Pythagoreans were concerned. To be sure, faint heart never won fair lady, and nowhere could this be more true than where Fama is concerned. But still, an aspiring mathematician might well pause on Hippasus' fate if he is pondering the advisability of establishing his reputation on the basis of an impossibility proof. For even in more enlightened times like the present the creator of an impossibility proof runs a special risk: one that further serves to highlight a key feature which characterises the significance of a mathematical theorem in general.

In one of his early confrontations with Turing, Wittgenstein insisted that 'All that the mathematical proof that a regular heptagon cannot be constructed with ruler and compasses achieves is to give us good grounds for excluding the phrase "construction of the heptagon" from our notation.'[17] As we shall see below the implications of this theme would be far-reaching, not only for one's attitude to impossibility proofs but much more importantly, for our understanding of the nature of

161

meta-mathematics and thence appreciation of the significance of Gödel's theorem. Let us ignore for the moment, however, the logical basis of Wittgenstein's argument and treat the above remark as a dogmatic thesis. One can well appreciate why mathematicians have objected so strongly to this passage; for it seems to deny the paramount role which Gauss' proof (that it is only possible to construct a regular polygon with an odd number of sides when it is either a prime Fermat number or a compound of such) played not just in Gauss' own life, but indeed, in the evolution of mathematics in the nineteenth century (*infra*). In Wittgenstein's defence it must be noted that the point he was trying to make here was solely concerned with the logico-grammatical significance of such a proof *vis-à-vis* the construction which it shows to be impossible. But even this is problematic, for Wittgenstein seems to hint that, seen in this light, an impossibility proof is of little more consequence than a warning that a certain construction is nonsensical. (This would appear to be the conclusion he emphasized when he continued: 'Hence "Smith drew the construction of the heptagon" is not false but meaningless.') But in that case, what sort of response could a mathematician expect from his peers for a demonstration that they had been squandering their efforts on an *unintelligible* issue; and indeed, is this not belied by Wittgenstein's very example?

According to Wittgenstein such proofs may continue to exert a strong historical fascination, but the chief result of this type of effort would be to close a chapter in the evolution of mathematical thought. To take a relatively minor example, a proof in chess theory that you cannot ø in *x* moves would be unlikely to evoke any strong responses. What it would accomplish would simply be to act as a deterrent: a chess master familiar with the proof would avoid landing in any such move-sequence which could at best end in stalemate. So too, viewed in the cold gaze of mathematical maturity an impossibility proof may strike us — apart from any value which it might have as an interesting exercise in the history of mathematics — as a warning of a dead-end which the prudent mathematician will avoid. And yet this was anything but the case with Gauss' proof; for among other things it played a central role in the evolution of group theory. Perhaps the greatest problem with this passage is thus the striking unsuitability of Wittgenstein's example for the purposes that he had in mind. But that does not rule out the point that he was trying to make, nor its possible bearing on his later remarks on Gödel's theorem, where the above remark on the significance of impossibility proofs was reproduced.

To understand Wittgenstein's intentions it is important to see that he was really thinking of the case where an impossibility proof only appears once the chapter in question has effectively been closed. For example, few today are familiar with the name of Pierre Wantzel even though he can justly be described as having proved the impossibility of one of the most famous of all construction problems: the trisection of an angle using ruler and compasses.[18] Here was a problem which had frustrated mathematicians for two millennia; surely its final proof should have earned its author everlasting reknown in the annals of posterity. But Wantzel's proof is distinctly treated by modern historians of mathematics as more of an anti-climax than a triumph. To be sure, it is always possible for mathematical tastes to become jaded; but the comparative insignificance of Wantzel's proof — as reflected by its present obscurity — seems to strengthen still further the case that the importance of a theorem is determined by the context in which it is produced and the influence which it thereby exerts.

To begin with, it is important that one bears in mind the manner in which interest in the trisection problem continually shifted. To the Greeks what mattered was ascertaining if, since it is possible to bisect an angle, it would also be possible to trisect an angle, whether this should be by using (in Pappus' terms) 'plane' methods (i.e. straight line and circle) or 'linear' (i.e. constructing other types of curves, such as Hippias of Ellis' quadratix). Had someone constructed (what was mistakenly regarded as) an impossibility proof of the former its chief effect would likely have been to intensify pursuit of the latter. To subsequent generations, however, the problem was no longer bound up with charting the methods for constructing regular polygons; what interest there was in the problem largely derived from the fact that their inability to provide a satisfactory explanation for *why* it is impossible to trisect an angle using ruler and compasses had become a source of frustration. So too it is crucial that we see how — as with the evolution of the calculus — the problem itself was constantly changing. The key to Wantzel's success lay in the creation of an algebraic *analogue* of the Greek problem. It is slightly misleading, therefore, to describe Wantzel's proof as a resolution of the *Greek* trisection

163

problem; indeed, as with the rigourization of the calculus, Wantzel's proof rests on the replacement of the geometrical framework which was the source of the Greeks' frustration with an algebraic approach (*infra*).

Without entering into the details of Wantzel's solution here,[19] it is worth remarking that the most important step in Wantzel's proof is the first, in which he established that construction problems using straight line and circle can be translated into algebraical terms and thence solved with a series of quadratic equations containing rational functions. From this premise he was able to demonstrate that the equation $x_3 + {}^3/_4 x + {}^1/_4 \alpha = 0$ (on which, as he had already shown, the algebraic analogue of the trisection of an angle depends) '*est irréductible si elle n'a pas de racine qui soit une fonction rationnelle de α et c'est ce qui arrive tant que a reste algébrique; ainsi le problème ne peut être résolu en général avec la règle et le compas.*' Restated in these algebraic terms the resolution of the trisection problem as it was now conceived was, as Wantzel went on to emphasize, immediately forthcoming. And herein lies the key to Wantzel's failure to make any impact on the history of mathematics. Had Wantzel's proof been instrumental in bringing this new algebraic framework to the attention of his peers his proof would undoubtedly have received considerable attention. But, of course, this was far from being the case. Rather, Wantzel's proof was formulated within the parameters of the exciting developments in modern algebra which had been initiated by the work of Ruffini and Abel. It was thus an immediate consequence of the great breakthroughs that had been achieved twenty years before. Indeed, Wantzel himself seems to have regarded his proof as little more than an off-shoot of the real issue at stake: the proof that the general equation of degree $n >$ 4 cannot be solved algebraically. Certainly there is no evidence to suggest that he had been led into the field of modern algebra by his desire to solve the trisection problem. It was only his efforts to improve on Ruffini's proof that the general quintic equation is unsolvable by radicals which eventually led Wantzel to apply the same methods in his demonstration of the impossibility of the trisection problem.[20]

Wantzel's proof may have been the final[21] step in the search to provide an adequate explanation for the impossibility of the trisection problem — thereby closing one of the most prolonged episodes in the history of mathematics — but it had only come about as a result of the developments in the theory of equations. It would be going too far to say that Wantzel's proof is pedestrian, but it is clear that the problem had become passé. There is obviously some truth, therefore, to Wittgenstein's pessimistic view of impossibility proofs (and it is perhaps for this reason that there is as yet no comprehensive work on the history of impossibility proofs *per se*). That is not to say, however, that it lies in the very nature of impossibility proofs that they can only perform this sort of subsidiary — 'closure' — role, riding on the crest of previous major advances. On the contrary, an impossibility proof can itself be the vehicle for the type of revolutionary change we have been considering. For in some cases, the key which fastens one lock may also serve to open another, in much the same way that a revolutionary theorem can open up an entirely new domain of mathematics.

No better example of this type of 'transitional' proof can be found than the work in modern algebra just alluded to which stemmed from the impossibility proofs carried out by Ruffini and Abel. Ruffini has also beeen passed over to some extent by modern historians of mathematics, but in his *Teoria generale delle equazione* he proved that no rational function of $n$ elements exists which would have 3 or 4 values under permutations of the $n$ elements when $n > 4$. This in itself was an accomplishment worthy of a passing mention in Rouse Ball's history. But Ruffini's main claim to fame hinges on the fact that in his — ultimately unsuccessful — attempt to prove in his *Riflessioni intorno alla soluzione delle equazioni algebraiche generali* that the general equation of degree $n > 4$ cannot be solved algebraically, he articulated what appears to be an obscure (and slightly muddled) precursor of Abel's theorem (*infra*). It is difficult to say how Ruffini would have been remembered had his proof succeeded, but it is nonetheless interesting to contrast his fate with that of Abel himself. For once again the point which this illustrates is that the significance of a theorem is not simply, as Hardy argued, a consequence of the ideas which it connects, but even more importantly, a product of the new domains which are thereby opened up.

There are obviously several factors involved in Abel's lasting fame, not least of which was his creation of a new branch of analysis: Abel's 'time outlasting monument', as Legendre described it. Nevertheless, Abel's influence on the development of modern algebra would have remained pronounced even had he never published any of his memoirs on elliptic functions. For in his solution of the quintic problem Abel hit on very much the same approach which Hilbert was to articulate at the end of the century. Indeed, the problem itself could be described as an early 19th-century candidate for Hilbert's 'Honours Class'; while Abel's attitude to his solution — and thus to mathematical problems in general — is remarkably similar to the main theme of Hilbert's lecture. In the introduction to *On the Algebraic Resolution of Equations* Abel explained that earlier algebrists had 'proposed to solve equations without knowing whether it was possible. In this way one might indeed arrive at a solution, although that was by no means certain; but if by ill luck the solution was impossible, one might seek it for an eternity, without finding it.'[22] This led Abel to formulate his own version of the general principle that 'in mathematics there is no *ignorabimus*':

> To arrive infallibly at something in this matter, we must therefore follow another road. We can give the problem such a form that it shall always be possible to solve it, as we can always do with any problem. Instead of asking for a relation of which it is not known whether it exists or not, we must ask whether such a relation is indeed possible. ... When a problem is posed in this way, the very statement contains the germ of the solution and indicates what road must be taken; and I believe there will be few instances where we shall fail to arrive at propositions of more or less importance, even when the complication of the calculations precludes a complete answer to the problem.[23]

Abel's first impulse had actually been to seek for an algebraic solution for the general fifth equation, which he initially thought he had discovered. Fortunately he realized his error before the paper could be published, which in turn led him to look at the problem from a completely different perspective: this time, to see if he could prove the impossibility of a solution. The first step of his proof was to establish the theorem that 'If an equation is algebraically solvable, then one can always give its root such a form that all algebraic functions of which it is composed, can be expressed by rational functions of the roots of the given equation.' He then went on to prove on the basis of this theorem the impossibility of solving the quintic equation by radicals. But in so doing he was not simply closing off one of the most well-known problems of his time, but was thereby creating a whole new range of problems and techniques in its place, whence group theory would subsequently evolve. For the full programme which Abel raised in the exposition of his proof was:

1. To find all the equations of any given degree which are solvable algebraically.
2. To determine whether a given equation is or is not solvable algebraically.[24]

Abel himself went some way towards answering this question (leading him to develop a partial classification of algebraic equations solvable by radicals, now referred to as 'Abelian equations'). It was then left to Galois to provide a complete characterization of the necessary and sufficient conditions for an algebraic equation to be solvable algebraically.

It was because of this stimulus that Abel's work motivated the development of modern algebra and group theory. The significance of a transitional impossibility proof is thus directly tied to whether or not it inaugurates a new field: and even more importantly, to the nature of the field which is thereby opened up. But then, quite a lot hinges on the manner in which such a proof is interpreted; and this in turn rests largely on the conceptual framework in which such proofs are received. How different would our attitudes be if Abel's solution had been viewed, not just as an impossibility proof, but in addition to that as an *ontological* or *epistemological* discovery? Here indeed would be food for metaphysical thought. Why was it that, in His infinite wisdom, God should have created algebraic solutions for general equations of the first four degrees, but not for the equation $ax^5 + bx^4 + cx^3 + dx^2 + ex + f = 0$? Is it the case that human powers are too limited to understand such a transcendent matter? Or have we simply not yet ascended to the 'meta-mathematical' level in which comprehension will be forthcoming? If Abel's proof was spared such conundrums Gödel's theorem unfortunately was not; a point which is itself of considerable importance for Wittgenstein's discussion. For while Gödel's theorem looks like — and was initially intended to be seen as — a *closure*, it has been widely interpreted as a *transitional* impossibility proof. But unlike the examples discussed above the revolutionary changes wrought by Gödel's theorem are not strictly confined to mathematical domains. And the chief purpose of Wittgenstein's puzzling remarks on Gödel's theorem was to make us aware of this anomaly, and eager to penetrate its secret.

## 2. ON THE SHORES OF METAPHYSICS

Mathematics has always skirted dangerously close to the shores of metaphysics. Inventor or Discoverer: the controversy over the most basic of mathematical concepts — the nature of mathematical proof — threatens to undo us. Yet for some reason, perhaps because of the undoubted respectability which they enjoy today, contemporary mathematicians remain untroubled by certain forms of platonist reflection. Even in logic we find that philosophers who would shudder at the thought of a transcendental deduction feel no compunction when it comes to discoursing on 'possible world' semantics. Of course, one explanation for these relaxed attitudes is simply that these modern species of metaphysics are not perceived as such; or else are viewed as scientifically sanitized versions. Certainly a good many of the philosophers who would be caught in the net of metaphysics as this was fashioned by Wittgenstein would be outraged at such an imputation. Yet the gap between intention and result can be enormous; particularly where the eradication of metaphysical speculation is concerned. For the great problem with such confusion is simply its subtlety; the secret to its abolition lies in the removal of the *premises* on which it rests, not the refutation or circumvention of its conclusions.[25] But whatever the explanation the fact remains that, for all the vaunted positivism of the modern scientific age, atavistic metaphysical instincts are never far beneath the surface.

Still, we seem to leave such disquieting matters far behind when we turn to the pristine axiomatic world of Gödel's incompleteness theorem. Or do we? Like Wantzel's resolution of the trisection problem Gödel's theorem results from an impossibility proof. But although Gödel's proof has all the earmarks of the closure variety the reaction which it has provoked is entirely consonant with that of a transitional impossibility proof. This may largely have to do with the death-blow which Gödel's theorem dealt to Hilbert's Programme and through that the deterrent influence which it has had on the foundations dispute. But then, is the foundations crisis — let alone Hilbert's Programme — still a vivid mathematical concern? Without underestimating the importance of Gödel's theorem in this area — or indeed, of the topic itself — it seems clear that the extraordinary impact which Gödel's theorem has had on contemporary thought must be found in some further quarter.

What we have to look at are not simply the ideas which Gödel's theorem connected (or rather, thwarted) but also, the new domains which it has served to open up. Moreover, it is at this point where an entirely new twist is introduced into the argument. For Wittgenstein would have us ask: what if there should be a philosophical problem with the issues thus created? We would have no choice but to work back from these problems to their origins, not just in Gödel's theorem, but even more importantly, in the conceptual framework in which Gödel construed his theorem. It is crucial, however, that we proceed cautiously here. If there is something wrong in the interpretation — as opposed to the construction — of Gödel's theorem, this is not a matter which can be passed off lightly; for the ramifications of a serious critique of this aspect of Gödel's theorem would be profoundly disturbing, as every mathematical logician is fully aware. But there is an enormous difference between questioning the significance as opposed to the coherence of Gödel's theorem. The preliminary issue that must be addressed before considering Wittgenstein's remarks, therefore, is simply: are there any signs in the developments that have been prompted by Gödel's theorem to suggest that something might be amiss in the reception of what is, perhaps, the most celebrated of all meta-mathematical achievements? And the first thing to notice is that there is certainly something *unusual* about the orthodox interpretation of Gödel's theorem, in so far as the whole debate which this has sparked off takes place in philosophy, not mathematics journals.

There have been numerous attempts to describe the philosophical significance of Gödel's theorem, and naturally enough, all take as their starting-point the impact which it had on Hilbert's Programme. But philosophers are energetic beings if nothing else, and from the narrow concerns of a highly technical and abstruse exercise in mathematical logic have arisen thriving industries in the philosophies of language and mind. But this already seems to present two separate ideas which, *prima facie*,

169

pull the theorem in different directions. For the fact is that in none of the examples of impossibility proofs so far examined would it make sense to speak of any *philosophical* significance: the only dimension we have been concerned with is the mathematical import of the theorem in question as this is determined by the ideas which the proof connects or the range of new questions and problem-solving techniques which the construction of a new system opens up. One possible exception would be Hippasus of Metapontum's legendary discovery; a case where it only makes sense to speak of any philosophical significance because of the Pythagoreans' metaphysical background.[26] It would be moving far too quickly to suggest at this stage that something similar might be operating in respect to Gödel's proof; particularly in light of the fact that the latter was originally hailed for its role in the elimination of metaphysics (*infra*). Nevertheless, what can be concluded here is that the move to debating the 'philosophical significance of a theorem' — without, it should be stressed, attracting any notice, let alone censure — must by its very nature indicate that an important shift of some sort in the conceptual background to Gödel's theorem has occurred; a transition which, at the very least, should be exposed and if possible, clarified.

When Gödel announced his theorem it was immediately seized on by the Logical Positivists as the key to consolidating their crusade against metaphysics. In his 'Autobiography' Carnap recalled how the Vienna Circle had broken with Wittgenstein over the question whether 'it is possible to speak about language and, in particular, about the structures of linguistic expressions'.[27] The problem was to avoid the metaphysics which, according to the *Tractatus*, would be the inevitable result of any attempt to develop, as Carnap proposed, a 'purely analytic theory of the structure of [a language's] expressions'.[28] The great breakthrough in Carnap's thought resulted from his discussions with Gödel and the publication in 1931 of the second incompleteness theorem. 'By use of Gödel's method' he set about to demonstrate how 'even the metalogic of the language could be arithmetized and formulated in the language itself'.[29] Interest in meta-mathematics had already been stimulated by the work of Hilbert and Tarski; but now Gödel's theorem opened up an entirely new vista: it suggested a means of extending the basic notions of meta-mathematics to the philosophy of language.

Thus, from the seeds of Gödel's theorem was born Carnap's theory of logical syntax.[30] But the campaign to construct a suitable metalanguage in which to formulate 'more precisely the philosophical problems in which [the Circle] were interested' soon ran into difficulties, not the least of which was Gödel's own attitude to these developments. According to Eckehart Köhler, Gödel persuaded Carnap to (re)introduce some platonist elements into his argument.[31] Whether the influence of Gödel's platonist sentiments on Carnap was as pronounced as Köhler suggests, it is clear that Gödel himself had embarked on a course which was ultimately to lead to his famous platonist *Credo* in 'What is Cantor's Continuum Problem?'.[32] This was all moving in very much the opposite direction from the constructivist version of logicism and quasi-scientific theory of logical syntax which Carnap envisaged. Significantly, Carnap's anti-metaphysical tool was soon to be employed for the very purposes which it had originally been designed to subvert.

One of the most disconcerting consequences of Gödel's theorem has been the influence which it has had on the apparently latent metaphysical tendencies which so many philosophers of mathematics still harbour. Indeed, Gödel's theorem has become one of the great moving forces behind the modern resurgence of platonism, and one suspects that this is principally the reason why Wittgenstein was led to attack what he regarded as the philosophical confusions embedded in the conceptual framework which underpins Gödel's interpretation of his theorem. But to the great annoyance of mathematical logicians everywhere Wittgenstein approached Gödel's theorem in the manner of someone setting about to dissect a fairy tale. The latter invariably begins with an invitation to suspend one's judgement of disbelief; a great part of their charm then lies in the author's ability to render a picture which seems to harmonise with a familiar aspect of life, or which creates some sort of consistent picture. But then, one must not suppose that such internal consistency 'is the criterion of truth and existence'.[33] Indeed, part of the appeal of Lewis Carroll's fantasies derives from the tension which he created between consistency versus intelligibility. What pleases is the development of the conflict (in much the same way that we enjoy the unfoldings of any artistic subject). This is a point which Hofstadter effectively brings out with his continuous comparisons of Gödel's theorem to Escher's etchings. Such a leitmotif plays a particularly useful role in

Hofstadter's appreciation of the genius of Gödel's theorem; indeed, if Wittgenstein is right, Hofstadter has here settled on the most apposite of metaphors.

All this is harmless enough. Where the danger lies is when someone is unable to discriminate between fairy tale and fact. To restore his senses we must bring him to see the story for what it is by clarifying the nature of the various premises which sustain it.[34] Such, at any rate, is the guiding spirit behind Wittgenstein's investigation into the framework in which Gödel's theorem was conceived. Those critics who rashly dismiss Wittgenstein's remarks as muddled technical confusions are chiefly those unwilling to consider the significance of this larger concern. It is a curious position for foundationalists to take; for whether right or wrong the results of such an approach can only be to deepen our understanding of some of the most basic concepts in the philosophy of mathematics. As a propaedeutic to Wittgenstein's discussion, however, we have still to address the basic question which confronts us in this section: viz. what has led philosophers to pursue the *philosophical* significance of Gödel's theorem in the first place? And here we do not want more of the metaphors which inspire *Gödel, Escher, Bach*; even prose is something of an intrusion, albeit rendered necessary by the flights of fancy in which so many commentators have indulged. But perhaps we are wrong to dismiss Hofstadter's analogies so brusquely?

Taking a leaf out of Hofstadter's own book we might consider why he strives to elucidate Gödel's theorem in his chosen terms: is it a coincidence that Hofstadter turns to Carroll and Escher in order to elucidate the meaning of Gödel's theorem? The nature of the task he has set himself is to clarify what Gödel's theorem means and how it operates: is there a hidden pressure for Hofstadter to turn to the sort of examples upon which he lights? If so it is precisely this internal dynamic that we must disclose. To be sure, there is a natural — perhaps essential — tendency to seek to interpret scientific or mathematical theories with metaphors; but it nonetheless bears remarking the sort of metaphors that Hofstadter has chosen. For he takes us through a catalogue of fictions all of which belong to the same category: they all violate the rules of logic in one way or another. But is the result of such transgressions the revelation of some hidden profundity as Hofstadter would have us accept? Like so many metaphysicians before him Hofstadter falls victim to the desire to transcend the bounds of sense. It is partly for this reason (and no doubt because of its popular tone) that *Gödel, Escher, Bach* has received so little attention from the mathematical community. But the most intriguing question posed by Hofstadter's book — which this silence only serves to intensify — is whether these paradoxes reveal some important feature of Gödel's theorem?

After all, why not simply elucidate the meaning and/or significance of Gödel's theorem in straightforward prose? One suspects that the answer to this question is to be found in the frequent (albeit covert) intrusion of the transcendental. Admittedly it is only near the end of the book that the term makes its one — brief — appearance; and significantly, it is promptly dismissed from the scene and replaced with the jargon, 'ism'.[35] For Hofstadter is embarrassed by the connotations of the term and substitutes in its stead an apparently secular neologism. But *Gödel, Escher, Bach* is a work which is obsessed with the transcendental. In a persistent effort to rehabilitate the concept Hofstadter turns first to Zen which is then purged of its religious overtones, leaving us with 'ism'. For in order to make sense of the illogical one needs an attitude remarkably akin to what would ordinarily be described as Faith.[36] It is clear from this why Hofstadter turns to Zen: it seeks to introduce a notion of meaninglessness which has its own special meaning (i.e. ineffable yet cognitively significant). Language cannot encompass the content of Zen; no words can express the truth of a Zen parable. For Zen transcends the bounds of sense: it escapes the 'narrow limits' laid down in Wittgenstein's later writings on the philosophy of language, and returns us to the 'real point' of the *Tractatus.*

Thus 'Ism is antiphilosophy: a way of being without thinking. Mumon penetrated into the Mystery of the Undecidable as clearly as anyone.'[37] A koan is then produced to illustrate this point; for koans construct object-sentences which are strictly unintelligible but which can be mapped onto meta-language propositions which express a truth. They are vivid examples, Hofstadter believes, of the limits of language *vis-à-vis* thought by hinting at the presence of higher truths. And this is precisely where the great appeal of Gödel's theorem lies in Hofstadter's eyes: in the apparent indication of how we can move *beyond* a system from *within* that system. He makes this connection explicit when he argues that, as in the case of Zen where truth surpasses the compass of words, so too in mathematics truth outstrips incomplete formal systems.[38] Just as the koans of Zen reveal the constraints on thought imposed by words — so that language itself becomes a barrier to understanding — so too Gödel's theorem established the existence of mathematical truths lying outside the ken of formal systems. Mathematical propositions cannot be circumscribed by formal systems: there will always be mathematical truths that lie beyond.[39]

By this time even die-hard platonists must be becoming distinctly uncomfortable, but do they have any right to disregard Hofstadter's excesses? For that matter, is this all that far removed from Gödel's own reaction to his theorem (*infra*)? Even those most actively opposed to the metaphysical flourishes in *Gödel, Escher, Bach* may be committed to the same premises which inspire Hofstadter's 'eternal braid'. In 'The Philosophical Significance of Gödel's Theorem' Michael Dummett sets out to refute the suggestion that Gödel's theorem undermines anti-realism on the grounds that such an objection operates on too crude a conception of this theory of meaning (and also because it assumes a spurious notion of a mathematical model).[40] But the problem with Dummett's reading no less than Hofstadter's 'metaphorical fugue on minds and machines in the spirit of Lewis Carroll' shines forth from its very title. At the outset of his argument Dummett states:

By GÖDEL'S THEOREM there exists, for any intuitively correct formal system for elementary arithmetic, a statement U expressible in the system but not provable in it, which not only is true but can be recognised by us to be true: the statement being of the form $(x)A(x)$ with $A(x)$ a decidable predicate. If this way of stating Gödel's theorem is legitimate, it follows that our notion of 'natural number', even as used in statements involving only one quantifier, cannot be fully expressed by means of any formal system. The difficulty is to assess precisely the epistemological significance of this result.[41]

But it is not so much the question whether Gödel's theorem can be used to defend (in Dummett's terms) a realist or anti-realist semantics but rather, what it means to speak of *epistemology* in the first place here.

The philosophical debate about the significance of Gödel's

theorem has been exactly this: a dispute over its epistemological consequences. But how does a *theorem* become the vehicle for epistemological doubts and discoveries? On such an approach we end up with a problem which, while it may be familiar to the philosophy of mind, strikes a discordant note as far as mathematics proper is concerned. Dummett mentions as a 'natural objection' to the 'meaning is use' thesis — which lends further conviction to the platonist interpretation of Gödel's theorem — the argument that

> since I cannot look into another man's mind in order to read there what meaning he attaches to 'natural number', since all I have to go on is the use which he makes of this expression, I can never know for certain that he attaches to it the same meaning as I do. Thus a place is left for scepticism of a kind closely related to that expressed by one who asks: 'How do I know that what we both call "blue" does not look to you as what we both call "red" looks to me?'[42]

But now the task is set — in exactly the same way as applies, not only to the foundations crisis but even more revealing, to such traditional philosophical problems as the worries over our knowledge of other minds — of refuting a looming sceptical attack. And one of the central themes throughout Wittgenstein's writings is that in all such cases our response must be the same: to scrutinize the premises which have led us into this dilemma.[43]

Before we dismiss this issue as the typical result of philosophical interference in non-philosophical matters it must immediately be stressed that this is very much the terms in which Gödel himself perceived the matter. In his autobiographical account of the development of his theorem[44] Gödel recalled how

> The completeness theorem, mathematically, is indeed an almost trivial consequence of Skolem['s 'Some Remarks on Axiomatized Set Theory']. However, the fact is that, at that time, nobody (including Skolem himself) drew this conclusion. ... This blindness (or prejudice, or whatever you may call it) of logicians is indeed surprising. But I think the explanation is not hard to find. It lies in a widespread lack, at

that time, of the required epistemological attitude toward metamathematics and toward non-finitary reasoning.[45]

As became clear in his work on the Continuum Hypothesis the attitude Gödel had in mind was thoroughly platonist. For he accepted as a consequence of his first incompleteness theorem the thesis that there is no obstacle to the notion of 'true but unprovable mathematical propositions'. Admittedly, there are legitimate cases in which we do speak of the existence of true but unproved mathematical expressions: viz. the axioms and postulates of a system. And given the epistemological obscurity surrounding these constructions it was consistent (or at least unsurprising) that the same problem should arise in regards to the unprovable 'propositions' entailed by these truths. Hence it seems straightforward enough to argue that the advent of Bolyai-Lobatchevskian geometry 'called attention in a most impressive way to the fact that a *proof* can be given of the *impossibility of proving* certain propositions within a given system'.[46] But we must proceed cautiously here, for it is one thing to speak of proving the impossibility of deriving the parallel axiom, but quite another to speak of proving the impossibility of *propositions* within a given system.

There is a crucial difference between independence proofs of the Parallel Postulate and e.g. Wantzel's proof of the impossibility of trisecting an angle using ruler and compasses: in the former it is shown that the axiom cannot be treated as a mathematical proposition in Euclidean geometry (i.e. a theorem in the system) whereas in the latter it was proved, not that a mathematical proposition is false but rather, that a certain mathematical expression is ill-formed. By disregarding the logical distinction present here one is led into the fallacy of treating mathematical conjectures as a form of hypothesis: i.e. meaningful propositions which have yet to be verified by a proof. But if the connection between mathematical propositions and their proofs is, as Wittgenstein insisted, *internal*, this can only mean that mathematical conjectures are ill-formed expressions themselves, albeit ones which may stimulate the creation of a set of rules which transform the conjecture into a proposition. Even this way of stating the matter is misleading, however; for in the absence of rules governing the use to which mathematical concepts are put an expression is condemned to remain ill-formed, and given that 'there are no gaps in mathematics', is condemned to remain a conjecture in its original context. In creating a new system we are, in effect, constructing an analogue of the conjecture which, in the new system, does indeed constitute a (meaningful) proposition which, by virtue of the Law of Excluded Middle, must be either true or false. For, far from having anything to do with epistemological issues, the Law of Excluded Middle is the basic logical criterion governing what we define as a 'mathematical proposition'. The question of how the mathematician arrives at mathematical truths quite simply does not arise here, therefore; Wittgenstein's sole concern was with clarifying the logical grammar of the concepts *mathematical proposition* and *conjecture*. This may not yet reveal too much about the substance of Wittgenstein's critique of Gödel's theorem, but at least it suggests where to look for an understanding of his motives.

## 3. WITTGENSTEIN'S INTENTIONS

Of the many problems posed by Wittgenstein's fragmentary remarks on Gödel's theorem perhaps the most frustrating is that it is not even clear how these fit into the overall context of Wittgenstein's proposed resolution of the foundations crisis. One of the first questions that must be asked, therefore, is simply: why should Wittgenstein have been so concerned with Gödel's proof in the first place? After all, this is an esoteric work which if anything should have contributed to the enhancement of Wittgenstein's own stated ambitions. Or at least so it has been argued, on the grounds that given the purpose of Gödel's theorem — to establish the impossibility of constructing a finitary consistency proof for arithmetic — it is far from clear why Wittgenstein should have sacrificed such an important potential ally. For the one thing that is manifest in Wittgenstein's writings on the philosophy of mathematics is his desire to challenge the two central themes that together constitute Hilbert's Programme: the formalization of mathematical thought, in order to produce a finitary proof of the reliability of mathematical reasoning.[47] But is the only explanation for this putative incongruity that — as so many critics have maintained — Wittgenstein failed to understand the purport and/or significance of Gödel's theorem?[48]

The chief drawback to such a reading — apart from the insult which it does to Wittgenstein's mastery of the subject — is that it ignores the basis for his attack on Hilbert's Programme. Only once this larger issue has been absorbed shall we be in a position to understand why and in what sense Wittgenstein objected to Gödel's theorem. For Wittgenstein's repudiation of the need for a consistency proof turned on the argument that Hilbert's Programme arises from a sceptical concern which is *ab initio* unintelligible: it introduces a spurious epistemological dimension into an issue which is solely concerned with the logical grammar of mathematical propositions. And since 'doubt can only exist where a question exists; a question can only exist where an answer exists, and this can only exist where something *can* be said' (NB 44),[49] the proper response to Hilbert's problem is to *dissolve* it by demonstrating that Hilbert's worry, far from being irrefutable, is *logically excluded* by the normative character of mathematical propositions.[50] Yet perhaps this leaves it doubly unclear why Wittgenstein should have troubled himself over Gödel's theorem, in so far as Gödel no less than Wittgenstein was intent on demonstrating the impossibility of satisfying Hilbert's Programme?[51] For that matter, why not talk of bypassing the latter, given the force of Gödel's results?

As with all impossibility proofs it is obviously crucial that we understand the conceptual background to Gödel's problem before assessing the significance of his results. In orthodox mathematical terms such a point goes without saying. For example, the whole interest of the Greek version of the trisection problem stemmed from the restrictions which were placed on the use of ruler and compasses in the proposed construction; remove these and, as we saw in the opening section, the problem is robbed of its original import. Likewise, it is essential that we fix the boundaries of the problem which occupied Gödel; otherwise there is a pronounced danger of either exaggerating or else distorting the significance of Gödel's theorem. In strictly meta-mathematical terms this point is no less straightforward; for if the restrictions which Hilbert imposed on the problem are relaxed — e.g. by permitting the use of transfinite induction — then it becomes possible (as Gentzen established in 1936) to prove the consistency of number theory.[52] But what if the issues raised by the framework conditions inspiring Gödel's interpretation of his theorem are *philosophical*, not mathematical; how then do we fix the boundaries of Gödel's problem? This is the question whence

Wittgenstein's critique of Gödel's theorem proceeds, rendering it so unlike any of the conventional approaches to Gödel's proof, and for that reason, so impenetrable to the traditional interests of mathematical logic.

In order to understand the strategy of Wittgenstein's critique it is crucial that we recognize how the epistemological foundation of Hilbert's framework accounts for the unique philosophico-mathematical character of Gödel's theorem. For Hilbert's Programme raises an issue which is *toto caelo* different from standard mathematical concerns; in Hilbert's words,

> The situation in which we presently find ourselves with respect to the paradoxes is in the long run intolerable. Just think: in mathematics, this paragon of reliability and truth, the very notions and inferences, as everyone learns, teaches, and uses them, lead to absurdities. And where else would reliability and truth be found if even mathematical thinking fails?[53]

As this passage makes clear, Hilbert was convinced that the paradoxes had induced an archetypal sceptical crisis which urgently demanded refutation.[54] And this epistemological reflex was in no way an anomaly on Hilbert's part.[55] As far as nineteenth-century mathematicians were concerned the immediate relevance of their work to philosophy — and vice versa — was obvious: mathematics was ensconced as the paradigm of *a priori* knowledge. It was a position which no one was prepared to abandon without a struggle. Thus, when the first serious blow to this stature was struck it was natural for mathematicians to respond *within* the framework of this conception. Gauss was moved to write, after discovering the possibility of constructing a consistent non-Euclidean geometry (which he was so reluctant to reveal 'for fear of the uproar of the Boeotians'):

> I am profoundly convinced that the theory of space occupies an entirely different position with regard to our knowledge *a priori* from that of [arithmetic]; that perfect conviction of the necessity and therefore the absolute truth which is characteristic of the latter is totally wanting in our knowledge of the former. We must confess, in all humility, that number is solely a product of our mind. Space, on the other hand,

179

possesses also a reality outside our mind, the laws of which we cannot fully prescribe *a priori*.[56]

If our knowledge of spatial reality is not *a priori* it must be *a posteriori* and the truths of Euclidean geometry must relinquish their claim to absolute necessity; for the framework decreed that there could be no third alternative. The collapse of geometrical certainty was the unavoidable result, therefore, of the classical epistemological conception of mathematical truth (as the discovery of hyper-complex numbers was soon to demonstrate). Moreover, any lingering faith in the reliability of 'geometrical intuition' was soon to be destroyed by Weierstrass' demonstration of continuous but nowhere differentiable functions.[57] The possibility that the framework itself might be to blame — that the *a priori/a posteriori* distinction might have nothing to do with the necessity of geometry — was never contemplated. And for all his mathematical radicalism Hilbert remained a philosophical conservative throughout his life, committed to what he saw as the task of salvaging the certainty of mathematical truth.

As noted above, the immediate significance of the second incompleteness theorem lay in Gödel's demonstration that given the mathematical parameters of the issue as laid down by Hilbert it would be impossible to construct a finitary consistency proof such as Hilbert envisaged. But this does not exhaust the matter: simply because the essence of Hilbert's Programme is its epistemological focus. And Gödel's theorem does not establish that Hilbert's Programme is unintelligible; only that Hilbert's goal is unattainable if one is confined to Hilbert's terms. It is all too easy to overlook the fact that Gödel's theorem only succeeds in subverting Hilbert's design by assuming from the outset that Hilbert's sceptical dilemma is cogent (and as far as Gödel was concerned, compelling). And the key to why Wittgenstein reacted so strongly to Gödel's theorem was precisely because it served to entrench even further the assumptions which had precipitated the foundations crisis. Thus Wittgenstein explained: 'It might justly be asked what importance Gödel's proof has for our work. For a piece of mathematics cannot solve problems of the sort that trouble *us*. — The answer is that the *situation*, into which such a proof brings us, is of interest to us. "What are we to say now?" — That is our theme' (RFM VII § 22).

Why, then, did Wittgenstein not devote greater attention to

the 'situation' into which Gödel's theorem brings us? Surely wherever sceptical illusions are allowed to breed they should be confronted and extirpated; especially when they are as influential as Gödel's theorem? On this line of attack Wittgenstein is accused, not of going too far, but of not going nearly far enough. For not only did Wittgenstein shirk his philosophical duty: he even confessed his desire — or at least, so it seems — to abandon it. No doubt could Wittgenstein have anticipated the extent of the interest that has developed in the alleged metaphysical consequences of Gödel's proof he would have turned to this task with vigour. But the real answer to this charge is that the brief critique of Gödel's theorem offered at VII §§22ff is not the premise but rather the conclusion of an argument that is developed throughout the course of *Remarks on the Foundations of Mathematics*. It is only by placing them in their proper context within the body of the work, therefore, that Wittgenstein's criticisms can be appreciated; for what Wittgenstein set out to establish was that the problem with Gödel's proof lies in the type of question which Hilbert was trying to answer 'meta-mathematically', and not in the complication which Gödel introduced into the issue. In one of his notebooks Wittgenstein remarked:

It is not Gödel's proof which interests me, but the possibility which Gödel makes us aware of through his discussion. Gödel's proof develops a difficulty which must appear in a much more elementary way. (And herein lies, it appears to me, Gödel's greater service to the philosophy of mathematics, and at the same time, the reason why it is not his particular proof which interests us.)[58]

It is the 'difficulty' which lies at the heart of the platonist interpretation of Gödel's theorem, and not the proof itself, which comes in for a minute philosophical analysis in *Remarks on the Foundations of Mathematics*.

With this in mind it is clear that Wittgenstein's reason for mentioning Gödel's theorem was because of the larger philosophical themes upon which it impinged. For as far as Wittgenstein was concerned the platonist interpretation of Gödel's proof feeds off a confusion which runs throughout the philosophy of mathematics. Gödel's conviction that he had demonstrated the existence of true but formally undecidable

mathematical propositions may be one of the most famous —
and for that reason influential — expressions of this confusion,
but it is itself merely a consequence of this deep-rooted
problem. Remove this core and the supposed metaphysical consequences of the theorem immediately collapse. Thus, perhaps
the most significant aspect of Wittgenstein's treatment of
Gödel's theorem is simply that it gets such short shrift; and to
some extent it is this desultory treatment which more than anything else has estranged mathematical logicians. For the hardest
part of Wittgenstein's attitude for them to accept is his veiled
suggestion that Gödel's theorem has little direct bearing on the
philosophical problems afflicting the foundations of mathematics.

In one sense, however, this interpretation underestimates
what Wittgenstein regarded as the profound implications of
Gödel's proof, although it is doubtful that disgruntled philosophers of mathematics will be any the more pleased with a
clearer understanding of the full dimensions of Wittgenstein's
thought. For Wittgenstein's objection is even more severe than
such an interpretation suggests: in Wittgenstein's eyes, whatever
significance Gödel's theorem possesses *vis-à-vis* the philosophy
of mathematics lies in its role, not as a *refutation,* but rather as a
*reductio ad absurdum* of Hilbert's Programme. If Gödel's
theorem does land us in a sceptical dilemma[59] then this can only
mean that the premises which initiated this line of thought must
themselves be *au fond* unintelligible. For 'Scepticism is *not* irrefutable, but *obvious nonsense* if it tries to doubt where no
question can be asked' (NB 44). Hence Wittgenstein's conclusion: 'I could say: it is not Gödel's proof which gives us the
stimulus to change the perspective from which we look at
mathematics. *What* he proves isn't what concerns us, but rather,
that we must come to grips with this *kind* of mathematical
*proof.*'[60]

The crux of this argument lies in Wittgenstein's insistence
that 'a piece of mathematics cannot solve problems of the sort
that trouble *us*'. There are two aspects to this claim; first, that
the type of problem *we* are concerned with (viz. the 'reliability'
of mathematical truth) is philosophical, and second, that
Gödel's theorem is confined to mathematics. This distinction
bears fundamentally on the epistemological approach to
Gödel's 'piece of mathematics'. Where platonists long to treat
Gödel's theorem as a new kind of transitional construction

Wittgenstein clearly regarded it as a closure impossibility proof. On the former reading Gödel's theorem takes us from the ruins of Hilbert's Programme into the sceptical depths or the meta-logical heights limned in the preceding sections. Whereas on Wittgenstein's reading Gödel's theorem closed off Hilbert's Programme: *full-stop*. The difference here rests on the demarcation between philosophical and mathematical problems. In order to treat Gödel's theorem in transitional terms this logico-grammatical distinction must be blurred, thereby allowing for the pivotal assumption that it is possible to solve philosophical problems mathematically. Thus the lead-up to Wittgenstein's remarks on Gödel's theorem is contained in the sustained examination of the logical barrier preventing the blending of philosophy and mathematics in the manner demanded by Hilbert's meta-mathematical recipe.

The heart of Wittgenstein's critique of the platonist interpretation of Gödel's theorem lies in the principle that it is only possible to resolve a philosophical problem *philosophically*; in this case by identifying and then removing the sources of the confusion which has led to the belief that there is a sceptical problem assailing the 'paragon of reliability and truth'. But once that task has been completed we are still left with a 'piece of mathematics' which — stripped of its metaphysical associations — lies waiting to be elucidated. And elucidated it must be if the lingering epistemological pull which shapes Hilbert's Programme is to be avoided, and thence, the platonist interpretation of Gödel's theorem. Wittgenstein wrote to Schlick: 'If you hear that someone has proved that there must be unprovable propositions in mathematics there is *nothing astonishing* about that because you have no idea as yet what this apparently clear prose proposition actually says. You have then to go through the proof from A to Z in order to see what it proves.'[61] But on Wittgenstein's account there are two independent issues which must be addressed in order to clarify what Gödel's proof actually proves: we must consider both the philosophical basis of Hilbert's problem and the mathematical content of Gödel's solution. For the fact that Hilbert's problem can only be resolved philosophically does not in the least entail that Gödel's proof is flawed.

It is this distinction which accounts for the otherwise perplexing consequence of Wittgenstein's argument that in one context Gödel's theorem is significant (albeit in a way that provides no succour for the platonist interpretation) while in another innocuous. For it is only if treated as a philosophical instrument that Gödel's proof serves as a *reductio ad absurdum* of the premise that philosophical problems can be solved mathematically. But in strictly mathematical terms there is no reason to deny that it is a legitimate piece of mathematics: which *per consequens* has none of the epistemological significance which the platonist interpretation contends. Once this philosophical issue has been dealt with it remains to elucidate the mathematical significance of Gödel's proof, and the problem reduces to the legitimate interpretation of Gödel's theorem. Like most issues in the philosophy of mathematics, therefore, the difficulty is one which only surfaces in the prose rather than the substance of the theory; in this case, in the divergence between Gödel's 'apparently clear prose proposition' and 'what his proof actually *proves*'.

This is by no means a straightforward matter, however, for the prose confusions which thrive in the interpretation of Gödel's theorem are buried deep in Hilbert's conception of meta-mathematics. The most notable feature of Hilbert's framework is that it landed him *in limine* in a similar problem to that which plagued logical positivist attempts to develop a tautological conception of mathematical truth. In 'Die Logischen Grundlagen der Mathematik' Hilbert outlined the basic themes of his proposed 'proof theory':

> Everything which constitutes mathematics today is rigorously formalized, so that it becomes a stack of formulas. ... To the ordinary, thus formalized mathematics, is added a, in a certain sense, new mathematics, a metamathematics. ... In this metamathematics one works with the proofs of ordinary mathematics, these latter themselves forming the object of investigation.[62]

But if the clarification — as opposed to the construction — of mathematical concepts belongs to philosophy rather than mathematics, this usurpation should encounter obstacles right from the start. And such has indeed been the case. Raymond Wilder asks: 'What is considered a *proof* in the metamathematics? For plainly the proofs embodied in that formal mathematics, which are to be justified in the mathematical study, must not be duplicated *in toto* in the latter! Otherwise we would be travelling in a vicious circle.'[63] There has been surprisingly little attention devoted to this issue. One way out of the impasse would be to

> Suppose, for example, that we could find a finite system of rules which enabled us to say whether any given formula was demonstrable or not. This system would embody a theorem of metamathematics. There is of course no such theorem, and this is very fortunate, since if there were we should have a mechanical set of rules for the solution of all mathematical problems, and our activities as mathematicians would come to an end.[64]

This recalls the objection which Poincaré raised against the logicist conception of mathematical truth.[65] The latter may seem a distant topic; but it illustrates the dilemma which confronted Hilbert in his attempt to justify his faith that 'There is no *ignorabimus* in mathematics'. For Hilbert had to defend his confidence in the solvability of all mathematical problems while at the same time avoiding the trivialization of mathematics.

Wittgenstein's response to this dilemma turns on the distinction between *mathematical questions* — whose meaning is determined by the rules of the system in which they reside — and *mathematical conjectures,* which by definition inhabit no system. Such is the platonist influence that even those most opposed to metaphysics find themselves drawn — unawares — to the notion of *discovering* the solution of unsolved problems. For they treat these questions as meaningful albeit lacking in the actual proof which will guarantee their truth or falsity. Admittedly the matter is less pronounced in constructivist than platonist writings. For the latter, proof is reduced (as can be seen in Hardy's *Apology*) to a trivial appendage introduced for the benefit of the incredulous or less gifted; whereas for the former proof constitutes the quiddity of mathematics. But there is a marked tendency for constructivists as well to treat unsolved mathematical problems as significant, even though such an assumption pulls them in opposite directions. On the one hand they maintain that the meaning of a mathematical proposition is determined by the logical connections which are articulated in a proof; on the other that mathematical conjectures are meaningful despite the fact that those logical connections have not yet been forged. But the upshot of their argument is that in the absence of a proof what one is dealing with is, strictly speaking, a heuristic device which guides the construction of a problem/solution. For to be a *meaningful* mathematical problem is *ipso facto* to be solvable, and it is for this logico-grammatical reason that 'There is no *ignorabimus* in mathematics.' There is thus an intimate connection between conjecture and theorem, but it is not one of synonymy.[66]

As important as Wittgenstein's proposed dissolution of the decision problem would be what principally concerns us here is the question of how meta-mathematics was supposed to circumvent this hazard. The solution which Hardy adopted was unsettling: meta-mathematics would be restricted to purely 'negative' results. In the passage from 'Mathematical Proof' quoted above Hardy continued:

Such a theorem is not to be expected or desired, but there are metamathematical theorems of a different kind which it is entirely reasonable to expect and which it is in fact Hilbert's dominating aim to prove. These are the negative theorems of the kind which I illustrated a moment ago; they assert, for example, in chess, that two knights cannot mate, or that some other combination of the pieces is impossible, in mathematics that certain theorems cannot be demonstrated, that certain combinations of symbols cannot occur.[67]

This is an arresting picture of meta-mathematics: one that does not so much move *beyond* as mutilate mathematics. Suppose, for example, that someone proposed to inaugurate the field of 'meta-geometry' with the sole proviso that the subject be confined to the investigation of impossible constructions.[68] Lest that does not sound worrying enough there would then be the further problem of clarifying the nature of the demarcation between ordinary and these 'meta-geometrical' impossibility proofs. Perhaps Wantzel's solution of the trisection problem actually belonged — unbeknownst to Wantzel himself — to this incipient 'meta-geometrical' discipline. After all, Wantzel's proof does not occur within the parameters of Euclidean geometry but rather, in the abstract realm of modern algebra. So is it the case that modern algebra is really a 'higher' geometrical system?

Such a possibility is by no means as remote as may at first

appear. For the notion of *meta-mathematics* did not spring Athena-like from Hilbert's mind unfertilized by previous developments. On the contrary it first made its appearance in Germany during the 1870s, where it was specifically employed in connection with the epistemological discussions that had been spawned by the development of non-Euclidean geometries. The question which preoccupied philosophers at the time was whether it is logically possible to conceive of beings inhabiting a four- (or higher) dimensional space; i.e. whether it is only a contingent fact that the space which we perceive is Euclidean. Thus the philosophical problems created by Bolyai, Lobatchevsky, and Riemann's proofs of the independence of the Parallel Postulate erupted during the 1870s into a metaphysical controversy over the bounds of spatial knowledge. The underlying worry in all this was whether mathematics could be sheltered from the collapse of geometry's claim to *a priori* certainty. Because of the nebulous character of these issues the term 'meta-mathematics' — introduced to describe these debates — initially took on a distinctly pejorative overtone; in Dühring's words, it was 'mathematical mysticism', and as such, to be avoided by practising mathematicians.[69]

Within a decade attitudes had begun to alter as interest in the study of the various conceivable types of physical space picked up. But it was Hilbert who played the instrumental role in this changing outlook.[70] Hilbert had a flair for taking ideas which had largely been neglected and giving them a new life by applying them in some unexpected fashion. In this case he made his influence felt in two different respects. To begin with he attacked the problem of space in the manner first broached by Pasch by treating the axioms of Euclidean geometry as generalizations about whatever satisfies them.[71] A set of axioms implicitly defines the primitive terms of its system and alternative (e.g. Euclidean versus non-Euclidean) geometries are not rival theories about space but rather alternative conceptual schema.[72] Then, as a signal of the way in which metaphysical problems could be removed by shifting one's point-of-view, Hilbert transformed the concept of meta-mathematics by making proof an object of mathematical study, thereby cementing the respectability which meta-mathematical investigations were beginning to enjoy. Without presuming at this point that there is a problem contained in these advances, it must at least be granted that they presage a profound philosophical shift. But the antecedents for this development date back — as the original meta-mathematical/geometrical discussions tacitly signified — to the work of Riemann, if not to the birth of analytic geometry two hundred years before.

In the first section it was asserted that Wantzel did not succeed in solving the Greek trisection problem but instead, that he created an algebraic version of the original problem which was readily solved in this new setting. The argument as dogmatically stated will no doubt strike many as exceedingly thin. After all, if it makes sense to say that Wantzel solved the geometrical trisection problem algebraically what reason could there be to prohibit describing this as the same problem which had interested the Greeks (as Wantzel himself clearly understood it)? Does this not follow from Vieta's discovery that classical geometrical problems could be addressed algebraically; or Descartes' creation of a coordinate method which could be used to solve classical geometric construction problems (as well as opening up an entirely new field of curves)? On Wittgenstein's approach, however, the trouble here starts with the temptation to suppose that Wantzel's proof establishes why the Greeks were unable to trisect an angle using ruler and compasses. What Wantzel's proof *actually proves* is that the algebraic analogue of the trisection problem does not satisfy an equation of degree $2^n$. Significantly, the terms of this solution would have been more mystifying to the Greeks than the original problem. For them to understand an impossibility proof of the problem it would have to be one that used the same methods and, what is all too easy to overlook, which understood the same notions by 'trisect' and 'angle'; i.e. it would have to be an impossibility proof *in their geometrical system*.

What tends to be forgotten in all this is that the Greeks *did* understand why the ancients were unable to trisect an angle: viz. the problem was not plane but solid and hence only solvable by conic sections. If one were to insist that what they failed to appreciate was precisely *why* this should be the case the answer is that it was solid because it was solvable by conics! What is perhaps most difficult for modern philosophers of mathematics to accept is that in their terms this just was the explanation of the matter; there was no epistemological problem forcing them to descend a step lower to the 'foundations' of the theory for this was the *bedrock* of their geometry.[73] The Ruffini/Abel/Wantzel solution operates in a foreign conceptual environment in which the parallels between the problems are sufficient to inspire the confusions that have been considered and yet not pronounced enough to generate widespread interest in Wantzel's proof. Moreover, it is interesting to speculate that with a few bold changes it might be possible to trisect an angle using ruler and compasses. To be sure, it might be a triangle the sum of whose angles $= \pi - \delta$ (where $\delta$ is a positive real number), or depend upon a flexible ruler such as Wittgenstein envisaged in Book I of *Remarks on the Foundations of Mathematics.* But then that is all the more reason to look judiciously at the relationship between Pappus' and Wantzel's approaches to the 'trisection problem'. For what we really have here might justly be described as a *family-resemblance problem*: a phenomenon which is perhaps unique to mathematics, and is primarily responsible for the grip which platonism exerts on the mathematical imagination.

This theme sheds further light on Wittgenstein's discussion of the mathematical significance of impossibility proofs. A naive response to an unproved expression would be to confess that one does not understand it. But such a frank admission would be met with scorn, even though its effect is the same as that of an impossibility proof by *reductio ad absurdum.* For mathematicians long to be certain that an expression is unprovable; until they have a proof it remains on the *ignorabimus* index to torment them. Wittgenstein demands, however, that we consider what sort of explanation a *reductio ad absurdum* provides. It literally demonstrates that if one takes the unproved expression and treats it as if it were a theorem a contradiction results. But what does *that* actually tell us, other than that there are no rules for the use of the expression in that system; which is exactly what our ingenuous mathematician had already indicated! Of course, there is a presumption that the impossibility proof establishes that the expression could *never* be proved. In regards to that particular system that must — per definition of 'impossibility proof' — be the case; but how many times in the history of mathematics have such problems — as the discussion of transitional impossibility proofs was intended to highlight — served as the locus for the creation of an entirely new system which takes as its starting-point the construction of a set of rules for the use of that expression?[74]

Where platonist confusions arise is when such novel proof systems are misconstrued as *extensions* of the preceding system which had proved 'inadequate' for the purposes in hand. If analytic geometry is viewed as an extended version of classical geometry and modern algebra as yet a further extension of the former the way is then open to treating Wantzel's proof as a meta-geometrical solution of the Greek trisection problem. On this picture it becomes compelling to see mathematics as an amorphous structure with successive generations ascending ever higher into the rarefied atmosphere of mathematical reality. And if this theory becomes plausible there is nothing to stop the platonist's projection of future upper strata: realms not yet disclosed, perhaps not even imagined, although some may be able to discern their faint outline. Here indeed emerge compelling grounds for a realist conception of mathematical truth; but only because these themes have been built into the theory as premises. If mathematical theories are seen as autonomous, interlocking systems which together make up the complex network of mathematics, however, this platonist illusion vanishes; and if we investigate, not the epistemological, but rather the logical status of a theorem or unproved expression, Hilbert's sceptical crisis correspondingly collapses, and in its place philosophers are left to investigate the *mathematical* significance of proofs. In any given case the elucidation of a theorem must proceed from and be strictly confined to the domains of the system in question. Or at least, it is on the basis of this theme that Wittgenstein's reading of Gödel's theorem evolved. But since Hilbert specifically introduced the notion of meta-mathematics in order to overcome this constriction that is obviously the next issue that must be addressed.


## 4. THE ROUTE TO META-MATHEMATICS: AXIOMATICS

It may seem unfair to lay the blame for the metaphysical excesses of some modern philosophers at the door of one of the greatest of all pure mathematicians. Yet it is one of the ironies in the history of philosophical ideas that someone as intent on eliminating scepticism and metaphysics as Hilbert should bear such heavy responsibility for the opposite result. For by removing the pejorative overtones which previously accrued to the notion of meta-mathematics, Hilbert opened up a vein of thought which has refused to stay confined to the narrow limits in mathematical logic which he envisaged. Perhaps Hilbert was guilty of hubris and the philosophy of mathematics has been left to confront the nemesis stalking his later foundational work. Not content with the major changes which he had instituted in the theories of invariants and algebraic number fields, Hilbert set out at the turn of the century to add a completely different type of creation to his list of accomplishments; not just another mathematical system, but a radically new *kind* of mathematics: *meta-mathematics.*

When recounting Hilbert's achievements Tarski (not unnaturally) reserved the development of meta-mathematics for special mention. Yet his praise was noticeably restrained:

> Hilbert will deservedly be called the father of metamathematics. For he is the one who created metamathematics as an independent being; he fought for its right to existence, backing it with his whole authority as a great mathematician. And he was the one who mapped out its future course and entrusted it with ambitions and important tasks. It is true that the baby did not fulfill all the expectations of the father, it did not grow up to be a child prodigy. But it developed sanely and healthily, it has become a normal member of the great mathematical family, and I do not think that the father has any reason to blush for his progeny.[75]

To be sure 'many mathematical developments are capricious in the extreme'[76] and seldom has an inquiry been called for to explain why a theory did not conform to the plan of its originator. This is a special case, however, for what is at stake here is not the manner in which meta-mathematics has been absorbed into 'the great mathematical family', but the reasons why its failure to live up to Hilbert's expectations has nurtured the very antithesis of his philosophical ambitions.

The issue has a particular importance for the present paper in light of the themes examined in the previous section; for this is the level at which Wittgenstein's criticisms of Gödel's theorem really operate. In *Philosophical Remarks* Wittgenstein objected:

> What is a proof of provability? It's different from the proof of proposition.
> And is a proof of provability perhaps the proof that a proposition makes sense? But then, such a proof would have to rest on *entirely different* principles from those on which

the proof of the proposition rests. There cannot be an hierarchy of proofs!

On the other hand there can't in any fundamental sense be such a thing as meta-mathematics. Everything must be of one type (or, what comes to the same thing, not of a type) (PR §153).

It may be tempting to try to disguise the scope of this attack; e.g. by showing how Wittgenstein's intention was not to impede the development of meta-mathematics but only to clarify in what the field consists. But the purport of this excerpt is that Hilbert's notion of a *supra-mathematical* system is untenable. The origin of Wittgenstein's approach to Gödel's theorem thus lies before the announcement of Gödel's results: a point of utmost importance which has hitherto been ignored. Indeed, the above passage is reproduced virtually unchanged in the context of the later remarks on Gödel's theorem, rendering the discussion of Gödel's proof at I Appendix I of *Remarks on the Foundations of Mathematics* an almost exact copy of the criticisms levelled against Hilbert's Programme at *Philosophical Remarks* §§152ff.

A typical example of the consequences of overlooking the basis of Wittgenstein's argument can be found in Alan Ross Anderson's review of *Remarks on the Foundations of Mathematics*. Anderson complains:

Wittgenstein proposes to shed general light on an important locution in meta-mathematics:

However queer it sounds, my task as far as concerns Gödel's proof seems merely to consist in making clear what such a proposition as: 'Suppose this could be proved' means in mathematics (p. 177).

But we are in point of clarity worse off after Wittgenstein's account than before.[77]

This passage reveals Anderson's lack of receptivity to what Wittgenstein said; for he disregards the wording of the quotation which he cites as an example of Wittgenstein's confusion. Whereas Wittgenstein declared his intention to clarify what Gödel's theorem 'means in mathematics' Anderson contends that he sought to explain 'an important locution in meta-mathematics'. But Wittgenstein's deliberate focus on the mathematical significance of Gödel's proof was grounded in the argument that, in so far as Hilbert's hierarchical conception of meta-mathematics is illicit — i.e. there are no meta-mathematical propositions in the manner conceived by Hilbert — Gödel's theorem has no *epistemological* significance and the coherence of Gödel's platonist gloss on the notion of mapping is dubious. For if we could perceive and describe isomorphisms between *higher* and *lower* systems then the route to a metaphysical interpretation of Gödel's theorem would indeed be unobstructed; as can be seen in Hofstadter's 'fugue' on an infinitely ascending hierarchy of meta-languages.

It was as a prophylactic against just such an interpretation that Wittgenstein warned:

it isn't enough to say that *p* is provable, what we must say is: provable according to a particular system.

Further, the proposition doesn't assert that *p* is provable in the system S, but in *its own* system, the system of *p*. That *p* belongs to the system S cannot be asserted, but must show itself.

You can't say *p* belongs to the system S; you can't ask which system *p* belongs to; you can't search for the system of *p*. Understanding *p* means understanding its system. If *p* appears to go over from one system into another, then *p* has, in reality, changed its sense (PR §153).

It is unfortunate that Wittgenstein should have couched his point here in the saying/showing terminology of the *Tractatus*. For it seems straightforward to state that *p* belongs to S (as Wittgenstein has just demonstrated). But Wittgenstein was not denying the possibility of speaking about *p* (e.g. in the prose of philosophy); only that you cannot say that *p* belongs to S *mathematically*, and cannot prove that *p* belongs to S without duplicating the axioms of S. This is a point which has already been touched on in the discussion of the status of Wantzel's proof *vis-à-vis* the Greek version of the trisection problem. The next step is to see why Wittgenstein felt this theme licensed the notorious declaration that his task was 'not to talk about (e.g.) Gödel's proof, but to by-pass it' (RFM VII §19).

The immediate problem with this avowal is that, as Wittgenstein feared, it does indeed sound queer; but its air of paradoxicality soon begins to dissipate once its aetiology is disclosed. The passage to which Anderson objects, for example, contains a veiled allusion to an argument first broached in *Philosophical Remarks* and developed at length in *Philosophical Grammar* which set out to undermine Hilbert's conception of meta-mathematics by clarifying the nature of mathematical propositions. The stress throughout is on *Hilbert's conception*: Wittgenstein repeatedly warned that while a philosopher may be an ombudsman he is not a legislator, and should not be tolerated in the larger role. It is not the place of philosophy to rule on whether or not meta-mathematics should be deemed illegitimate; philosophy's duty is to describe, not reconstruct. Hence Wittgenstein saw his brief as that of illuminating what meta-mathematics as treated by Hilbert involves: i.e. of separating the philosophical from the mathematical strands intertwined in Hilbert's Programme. That is not to say that neither of these threads cannot — if taken on their own — be labelled 'meta-mathematics'. But meta-mathematics as mathematically conceived must be stripped of Hilbert's notion of superjacent systems; while in philosophical terms Wittgenstein's argument might itself be described as meta-mathematical, if all that is meant by that is discussing the nature of mathematical concepts. We would have still to explain how the latter activity differs from the philosophy of mathematics — or rather, from philosophy *simpliciter* — but the real question is how it differs from *mathematics*. For the dangers present here all revolve upon the attempt to translate problems from the philosophy of mathematics into the province of mathematics proper.[78]

In order to grasp how this idea first gained currency we need to consider both the factors which conditioned Hilbert to the possibility of a mathematical hierarchy and the pressures which forced him in this direction. In *Philosophical Remarks* Wittgenstein indicated that the first step on the road to Hilbert's Programme lay in Hilbert's bid 'to establish for geometry a *complete*, and *as simple as possible*, set of axioms and to deduce from them the most important geometric theorems in such a way that the meaning of the various groups of axioms, as well as the significance of the conclusions that can be drawn from the individual axioms, come to light.'[79] As Hilbert here made clear *Foundations of Geometry* must be seen against the long-standing interest in 'identifying and filling the gaps' in Euclid's proofs.[80] It had been argued since ancient times that many of

Euclid's theorems depend upon unstated assumptions. A rigourous reconstruction of Euclid's deductive system would be one which rendered all the tacit premises fully explicit. But then, given the success of *Foundations of Geometry,* does not Hilbert's axiomatization of Euclidean geometry provide an obvious counter-example to Wittgenstein's insistence that 'there are no gaps in mathematical systems': an argument which Wittgenstein first formulated with direct appeal to the example of Euclidean geometry?[81]

To see how Wittgenstein's argument was designed to meet this objection we must distinguish between e.g. Weyl's affirmation that 'Euclid's list of axioms was still far from being complete; Hilbert's list is complete and there are no gaps in the deductions'[82] versus Wittgestein's concern with the notion of 'gaps in a system'. A case in point can be found in the first proposition in the *Elements.* Euclid set out to construct an equilateral triangle on the basis of a finite straight line. From a constructivist point of view the flaw in Euclid's proof lies in the third of his initial premises that 'from the point C the circles cut one another';[83] without an additional Principle of Continuity this assumption is unwarranted. Thus Heath remarked that 'It is sufficient for the purpose of this proposition and of I.22 where there is a similar tacit assumption, to use the form of postulate suggested by Killing.'[84] But the need for such a principle hardly entails that it is — tacitly — present in Euclid's proof. To suppose otherwise is to fall victim to the confusions that Euclid *implicitly* assumed Killing's postulate or that Killing merely articulated Euclid's 'hidden premise' (in terms which would have been incomprehensible to Euclid). But if not this exactly what was it that, in Torretti's words, was 'tacitly understood' by Euclid?

If the proof cannot proceed unless some such postulate is introduced this constitutes not a 'gap in Euclid's system' but a gap in his deduction, induced by his reliance on line drawings and hence no proof if the drawings are divorced from the formal body of the proof. If there is no method of demonstrating that the circles will intersect at point C then strictly speaking there is no sense to such a mathematical assertion; it is an 'idly turning wheel' in Euclid's 'proof'. But when a Principle of Continuity is introduced one has not thereby filled a gap in Euclid's system; rather, Euclidean geometry has itself been extended. And that means, not that a higher — perhaps 'meta-

Euclidean' — geometry has been developed into which the earlier proof has been transposed[85] but instead, that a more comprehensive system has been constructed in which expressions that were unintelligible in the previous system are now endowed with sense: 'We have in fact moved on to a new system that does not contain the old one but contains a part with exactly the same structure as the old system' (WWK 35-6). In Wittgenstein's terms 'Euclidean geometry' is itself a family-resemblance concept, and in such an addition to the chain of Euclidean geometries an analogue of the original *apodeixis* emerges as — by modern standards — a satisfactory proof.

That is not to say that it is impossible to restructure an existing system; an obvious example would be the proof that an axiom is superfluous.[86] A more contentious issue is the construction of what Nicod described as 'equivalent' systems (in which the totality of propositions remains th same but the status of axioms versus theorems is altered). On Blanché's reading 'all the axiomatic reconstructions of Euclidean geometry are equivalent since they contain, basically, the same set of terms and propositions: the difference is simply in the way in which the latter are divided up into primitive and derived.'[87] But on Wittgenstein's approach we would have to be far more scrupulous, given the definitional role played by axioms, in the ascription of propositional identity; an argument which in fact accords closely with Nicod's intentions. In 'The Formal Relationship Between Various Geometric Systems' Nicod warned: 'These propositions which appear to be the same in all systems, when the hierarchy alone is changed, are they truly identical? Let us examine this closely: their identity is only apparent.' For a 'system knows its own primitive expressions only and is incapable of discussing anything else. If we succeed in finding the same statements here and there, that is thanks to the device of identical abbreviations which conceal different expressions, and this is merely a clever play on words.' This means that the 'propositional identity' to which Blanché refers is

merely verbal. But it makes manifest a certain formal correspondence. It shows that it is *possible* to 'translate' the simple expressions of a system into compound expressions of another in such a way that the axioms of the first system (and consequently all its propositions) become translated into propositions of the second system. This is the relationship which logically binds together all conceivable systems of euclidean geometry.[88]

Wittgenstein did not so much depart from as refine Nicod's inception. The key element in his advance is the clarification of this notion of 'translation' as it applies not just to the restructuring of an existing but also to the creation of a new system which supersedes a (possibly non-) axiomatic theory. In the former case rules are stipulated for mapping the propositions from one system onto another. In the latter new axioms are introduced and the primitive concepts are subtly but significantly altered; a point to which Hilbert was also committed, for as he wrote to Frege: 'Every axiom contributes something to the definition [of a basic concept] and hence every new axiom changes the concept.'[89] It is precisely this theme which underlies Wittgenstein's remarks on completeness: 'The edifice of rules must be *complete*, if we are to work with a concept at all — *we cannot make any discoveries in syntax.* — For, only the group of rules *defines* the sense of our signs, and any alteration (e.g. supplementation) of the rules means an alteration of the sense. Just as we can't alter the marks of a concept without altering the concept itself. (Frege)' (PR §154). Mathematicians may speak freely of proving old theorems in an expanded or restructured theory, but the fallacy in this assumption is betrayed when proofs cannot be exchanged between two systems (as can be seen in the above example: the point of Killing's postulate just is that the theorem cannot be derived in a system which dispenses with line drawings but lacks a Principle of Continuity). In all such cases we must be careful to speak of mapping parallel propositions and proofs from two autonomous members of a family-axiomatic system onto one another.[90]

Far from denying, Wittgenstein's argument exploits the inevitable exhaustion of a system. Axioms are constantly being invented or revised in order to enable mathematicians to address problems which would defy the reach — i.e. the intelligibility — of the pre-existing axioms. The evolution of a mathematical field typically consists, therefore, in the gradual development of such system-families. But the further this process advances the greater become the pressures for rationalization if the subject is to remain manageable. Like the unplanned development of a medieval city these systems soon become too cumbersome for efficient communication, and interest shifts accordingly to refashioning the axiom set without diminishing the scope of the system (by introducing alternative — possibly more abstract — axioms). It is to this phase that foundational studies belong; like a modern urban planner engaged in imposing a gridwork on a renaissance city by removing cul-de-sacs and creating new trunk lines, the foundationalist sets out to build an efficient structure on the edifice of an old network.[91] The basic misconception underlying these investigations, however — highlighted by the fact that they are referred to as 'foundational' — is that they establish an *ultimately* complete system which brings the evolution of the system/family to its conceptual apex (cf. Weyl's comment *supra*).[92] For the earliest deductive systems were just as 'gapless' as these later versions. To be sure the latter may be more rigorous (and to modern tastes, more elegant) than their classical forbears, but that is a far different matter from the intrusion of the 'completeness' assumption.

The argument leading Wittgenstein to this conception of mathematical decidability bears an even more striking affinity to Hilbert's account of the solvability of any mathematical problem than that noted above. Hilbert cautioned in 'Mathematical Problems' that 'he who seeks for methods without having a definite [*bestimmte*] problem in mind seeks for the most part in vain.'[93] A 'well-defined' problem, according to Hilbert, is one whose presuppositions have been 'rigorously' formulated. When this proves impossible it is advisable to ascertain whether the problem 'is a special case of a more general problem', and if this too meets with failure we must consider whether we 'seek the solution under insufficient hypotheses or in an incorrect sense'.[94] Wittgenstein's distinction between *gaps in a deduction* and *gaps in a system* proceeds from this last point. The crux of Wittgenstein's argument is that whatever 'well-defined problem' can be asked can in principle be answered; but it is for that reason that any mathematical system, however primitive, is *ipso facto* complete. For 'Where you can't look for an answer, you can't ask either, and that means: Where there's no logical method for finding a solution, the question doesn't make sense either' (PR §149). To be a well defined — i.e. meaningful — *mathematical problem* is by its very nature to adhere to the rules of a system:

Mathematics cannot be incomplete; any more than a *sense*

can be incomplete. Whatever I can understand, I must completely understand. This ties up with the fact that my
language is in order just as it stands, and that logical analysis
does not have to add anything to the sense present in my
propositions in order to arrive at complete clarity. (PR §158)

That is not to deny that *Foundations of Geometry* is an
axiomatization of Euclidean geometry: provided the emphasis
on Euclidean geometry's being a family-resemblance concept is
understood. But it still leaves open the pertinence of the book's
title in light of its contents. Hilbert clearly saw the enterprise as
*foundational* in so far as it endeavoured to establish a 'complete, and as simple as possible, set of axioms' for Euclidean
geometry. But he also stated that he hoped 'to deduce from [his
five axiom groups] the most important geometric theorems in
such a way that the meaning of the various groups of axioms, as
well as the significance of the conclusions that can be drawn
from the individual axioms, come to light'.[95] There is a critical
difference between these two conditions, however; for whereas
the former might be described as a trivial aspect of any
deductive system the latter does indeed provide the moving
force behind Hilbert's enterprise. For what Hilbert constructed
was an axiomatic system in the algebra of three real variables
that can be mapped onto a large part of the Euclidean family.
Where Euclidean geometry had become a warren of deductive
systems (brought about by the postulates introduced over the
centuries) Hilbert created a global system which generates *ana-
logues*[96] of the major Euclidean theorems without having to
introduce intermediary postulates. But this did not show that
Euclid implicitly foresaw this global system; it demonstrated
how Euclidean geometry as it existed at the end of the
nineteenth-century could be mapped onto Hilbert's abstract
structure. Torretti claims that 'Every Euclidean theorem is a
logical consequence of Hilbert's axioms. The latter can therefore be said to provide an exhausitive [sic] conceptual analysis
of the object of Euclidean geometry.'[97] If the first part of this
argument is true it can only be on the understanding that
Hilbert's system marks an extension of the Euclidean family.
Hence the latter part of Torretti's argument can easily be misconstrued, for the only theorems which follow from Hilbert's
axiom groups are *those which are intelligible in Hilbert's
system*: a point which denies much of the inspiration behind the

confusing thought that a set of axioms (which are supposedly 'generalizations' about whatever satisfies them; *infra*) could provide a 'logical analysis of our perception of space'.[98]

In all fairness to Hilbert he did come close to succeeding in his philosophical task: not, that is, in *providing* but rather in *clarifying* in what the foundations of geometry consist.[99] But because he did not see the issue in these terms he failed to grasp the significance of his own results. When Frege objected (not unfairly given Hilbert's reference to the *meaning* of the axioms in his introduction, *supra*) that 'the axioms are made to carry a burden that belongs to definitions. To me this seems to obliterate the dividing line between definitions and axioms in a dubious manner'[100] (e.g. by suggesting that the axioms of a system partially define the primitive concepts and that an interpretation completes the process[101]). Hilbert responded:

> you say that my concepts, e.g., 'point', 'between', are not unequivocally fixed: 'Between' is understood differently on p. 20, and a point there is a pair of points. But it is surely obvious that every theory is only a scaffolding (schema) of concepts together with their necessary connections, and that the basic elements can be thought of in any way one likes. E.g., instead of points, think of a system of love, law, chimney-sweeps ... which satisfies all axioms; then Pythagoras' theorem also applies to these things. Any theory can always be applied to infinitely many systems of basic elements.[102]

Frege was aware that Hilbert sought 'to detach geometry entirely from spatial intuition and to turn it into a purely logical science like arithmetic.'[103] In a fundamental sense Hilbert had shown that geometry, like arithmetic, constructs rules which fix the use of the mathematical concepts employed in the description of reality. Just as arithmetical rules can be applied in myriad ways so too the 'points' etc. of geometry can be applied to 'infinitely many systems of basic elements'.[104] But as the correspondence with Frege makes clear, Hilbert thought that the concepts defined by the axioms of a system are without content until given an interpretation: i.e. pure geometry defines and applied geometries determine the meaning of primitive concepts. Hence we must be careful not to credit Hilbert with Wittgenstein's insight that 'the axioms of geometry have the character of stipulations concerning the language in which we want to describe spatial objects. They are rules of syntax. The rules of syntax are not about anything; they are laid down by us' (WWK 62). For where Hilbert maintained that pure geometry lays down the logical form of primitive concepts without understanding what they mean, thereby confusing the application with an 'interpretation' of these concepts,[105] Wittgenstein explained that the *meaning* of primitive mathematical concepts is constituted by the rules governing their use. Thus, where Hilbert regarded pure and applied geometry as completely independent of one another, Wittgenstein responded that the essence of geometrical as well as arithmetical concepts is that they *appear in mufti*: just as $2 + 2 = 4$ whether we are adding pebbles or people, so too all the points on the circumference of a Euclidean circle must be equidistant from the centre whether we are measuring the spatial configurations of tables, chairs, or mugs.

To say that *Foundations of Geometry* is an axiomatization of Euclidean geometry in no way licenses, therefore, the much stronger conclusion that it is a paradigm of 'formal axiomatics'. The *Foundations* remains a notable mathematical achievement (which by definition is one of concept-construction, not analysis). Its logical relations are more rigorously (if not more perspicuously) developed than is the case in earlier contributions to the evolution of Euclidean geometry. But precision in itself may seem a poor reward for so mammoth an undertaking, and Hilbert certainly felt he had achieved far more. After all, this was a work on the *foundations* of geometry; what Hilbert thought he had demonstrated was the reliability of Euclidean geometry. For if you could establish the consistency of the axioms — in whatever form these might take — you would have ensured that Euclidean geometry provides a true representation of space.[106] To be sure, the same could be done for Bolyai–Lobatchevskian geometry, in which case Hilbert had actually established that there are any number of — in his terms — 'true' geometries. But Hilbert was not intent on rescuing the claims of Euclidean *vis-à-vis* non-Euclidean geometries, nor even in refurbishing the reputation of Euclidean geometry. His aim was to show that the creation of non-Euclidean geometries did not pose any threat to mathematical truth *per se*: provided the consistency of Euclidean geometry could be assured.

The point of Hilbert's relative consistency proof of Euclidean geometry in the theory of real numbers was thus to show that the former has as much claim to truth as the latter.[107] Unfortunately the argument placed Hilbert on a slippery slope (an alarmingly common phenomenon in foundational studies). For the price of shoring up the certainty of Euclidean geometry in this manner was — following the announcement of the discovery of the paradoxes in set theory — to expose that of higher mathematics *simpliciter*. Ironically, where a large part of Hilbert's reasons for writing the *Foundations* appears to have been to refute Kronecker's exclusion of geometry from pure mathematics, he only succeeded in — temporarily — salvaging the truth of geometry at the price of undermining that of pure mathematics itself. The problems contained in Hilbert's initial attempt to provide an axiomatic foundation for geometry were thus passed on to his meta-mathematical attempt to prove the absolute consistency of arithmetic. But 'passed on' is not strong enough, for the course on which Hilbert had embarked led straight into meta-mathematics. To see this we must turn to the Hilbert-Poincaré debate on the role of intuition in geometry and arithmetic.

## 5. THE NATURE OF META-MATHEMATICS: WITTGENSTEIN'S STANDPOINT

Poincaré's response to *Foundations of Geometry* was not entirely encouraging:

> Attempts have been made, from another point of view, to enumerate the axioms and postulates more or less concealed which form the foundation of different mathematical theories. ... It seems at first that this domain must be strictly limited, and that there will be nothing more to do when the inventory has been completed, which cannot be long. But when everything has been enumerated, there will be many ways of classifying it all. A good librarian always finds work to do, and each new classification will be instructive for the philosopher.[108]

The sting in the tail of this qualified support is clear: interesting as these bibliographical exercises might be, the 'genesis of mathematical knowledge' remained in the balance. Hilbert may not have intended to dispute the role of intuition in geometry (as the Kantian epigram was intended to bring out)[109] but the conventionalist framework of the *Foundations* left no room for the putatively synthetic *a priori* truths delivered by intuition. Thus Poincaré objected that

> The logical point of view alone appears to interest Professor Hilbert. Being given a sequence of propositions, he finds that all follow logically from the first. With the foundation of this first proposition, with its psychological origin, he does not concern himself. ... The axioms are postulated; we do not know from whence they come; it is then as easy to postulate A as C.[110]

Poincaré had seized on the very point which, following his correspondence with Frege, Hilbert was to focus on in his approach to the foundations crisis. Where Poincaré struggled to explain the genesis of mathematical knowledge in terms of *Urintuition*, Hilbert turned to meta-mathematics. In his 1904 response to the discovery of the antinomies in set theory, 'On the Foundations of Logic and Arithmetic', Hilbert insisted that it must be possible to develop a 'rigorous and completely satisfying foundation' for the concept of number, encompassing (contra Kronecker) the irrationals as well as rationals. The paper adheres to the opaque tradition in which '1', '0' and '$\aleph_0$' are defined as 'thought-objects' ( *Gedankending* ). But it presents two revolutionary ideas: first, that if mathematics could be 'formalized' the consistency problem could be reduced to the derivability of a formula in a formal system expressing both a statement and its negation. And second, that a consistency proof could serve as a demonstration that a property possessed by the axioms (which Hilbert characterized as 'homogeneousness') is passed on by the rules of the system to all of the theorems.

The following year Poincaré returned to the attack. The trouble with trying to secure the foundations of mathematics with consistency proofs for formal systems is that these must refer to *all* proofs in the system; since they must employ they cannot then be used to justify mathematical induction.[111] Poincaré's criticism seems to have found its mark, for it was not until 1922 that Hilbert returned to the issue. He accused

Poincaré of precipitately rejecting the possibility of a consistency proof for arithmetic because he believed that the consistency of induction could only be proved via mathematical induction. He was led into this assumption because he had not recognized the importance of the mathematics/meta-mathematics distinction, and thus failed to see that what he regarded as mathematical induction intuitively known is a completely different process.[112] Hilbert qualified this position in 1927, maintaining that there are two principle uses of induction: *contentual* and *formal*;[113] instead of using mathematical induction in meta-mathematical proofs we appeal to paradigmatic examples.[114] He now castigated Poincaré's version of mathematical induction as one of the spurious 'transfinite' methods of inference, insisting that meta-mathematics deals with concretely given objects to which only finitary recursive methods apply. Poincaré's objection is rejected on the grounds that, as Hilbert subsequently explained in 'Die logischen Grundlagen der Mathematik', arguments containing 'all' belong to the transfinite.[115]

When he turned to the foundations of mathematics Hilbert found that his problem was now two-fold, as forced on him by Poincaré's frequent criticisms of axiomatics.[116] His priority remained to eliminate scepticism from mathematics, but he had to do so in such a way that the possibility of future discoveries would not be jeopardized. The way out of this dilemma was to be found in meta-mathematics: the truth of the axioms of a given system would be established by a meta-mathematical consistency proof and the *interplay* between object and meta-mathematical systems accounted for the development of mathematical thought.[117] By 'interplay' (*Wechselspiel*) Hilbert had in mind a dialectical process in which an object system provides the raw material (contentual propositions) which are formalized in a meta-system, yielding 'ideal propositions' which might be applied in unexpected domains.[118] The formal system itself, far from being confined to merely (re)classifying the elements of a theory, could then provide the source of important new advances.[119] (It was chiefly for this reason that Gödel's second incompleteness theorem could still be greeted enthusiastically by Bernays; for although the sceptical problem might have been revived Gödel's proof also seemed to vindicate Hilbert's view of the evolution of mathematical theories.)

It might at first appear that Hilbert only landed himself in a fresh dilemma with this argument, in so far as he had now to account for the development of mathematics prior to the inception of meta-mathematics. Yet he could argue that the tension between 'form' and 'content' which underpins his conception of the genesis of mathematical/scientific knowledge had always been present albeit in a different guise.[120] The introduction of formal systems had been rendered necessary by the discovery of the paradoxes, and was licensed by the fact that meta-mathematics slips neatly into place in the form/content bifurcation, enabling us to resolve the foundations crisis while satisfying Poincaré's qualms about mathematical discovery. Two different problems arise with this argument, however: first, whether Hilbert's identification of meta-mathematics within the bosom of formal studies is warranted; and second, whether Hilbert correctly interpreted the tension between these two aspects of mathematical creation. For it is important to be clear that the axiomatization is not the same thing as the formalization of a system; the former may be a necessary condition of the latter but the reverse is certainly not the case.

Bourbaki are scrupulous about this point. In the introduction to *Theory of Sets* they explain that

The axiomatic method is, strictly speaking, nothing but [the] art of drawing up texts whose formalization is straightforward in principle. As such it is not a new invention; but its systematic use as an instrument of discovery is one of the original features of contemporary mathematics. ... Just as the art of speaking a language correctly precedes the invention of grammar, so the axiomatic method had been practised long before the invention of formalized languages.[121]

In 'The Architecture of Mathematics' they warn that 'The words "formalism" and "formalistic method" are ... often used; but it is important to be on one's guard from the start against the confusion which may be caused by the use of these ill-defined words, and which is but too frequently made use of by the opponents of the axiomatic method.'[122] Their final verdict is circumspect: the 'formalistic method is but one aspect of [axiomatics], indeed the least interesting one'.[123]

Wittgenstein demanded that we go a step further; not only is it unnecessary to formalize a system in order to axiomatize it: it is impossible, in the sense in which this is *mathematically* understood. For as Frege had demonstrated, the two concepts are incompatible; the deductive notions operating in the latter are unintelligible in the context of the former, making it senseless to speak of *deriving* — as opposed to *generating* — one string of meaningless marks from another. This crucial point is completely masked over by the formalist distinction between 'material' and 'formal' axiomatics. That Bourbaki did not draw this conclusion is explained by their conception of formalism: 'By analysis of the mechanism of proofs in suitably chosen mathematical texts it has been possible to discern the structure underlying both vocabulary and syntax.'[124] On this (orthodox) reading formalism *strips away* the semantic content leaving the syntactical structure of a proof exposed. But that merely assumes that it makes sense to apply the term 'proof' to transformations of symbols according to canonical rules. The heart of the issue thus lies in the formalist premise that mappings assign a meaning to syntactic formulae.

The prominent schools in the foundations dispute had all agreed that mathematical propositions are *referential*; where they differed was on whether the objects denoted are physical, mental, or abstract. Hilbert hoped to escape from this ontological quandary by first isolating the 'scaffolding' of mathematical propositions from their referential character and then bridging the resulting gap between syntax and semantics with 'interpretations'. But while Hilbert had indicated that any set of axioms could be treated as a system of conventions, he failed to explain how and why these conventions should be regarded as *mathematical.* For the result of his conception is to render the connection between pure and applied mathematics *external*: we not only can but indeed must speak of 'pure mathematical systems' independent of and prior to their applications. In which case the problem is to explain in what sense 'meaningless marks on paper' can be described in mathematical terms: i.e. how the family of deductive concepts can be applied to formal calculi. For all of the basic concepts have been redefined so as to strip them of their inferential content. As Hofstadter explains,

> The term 'theorem' has, of course, a common usage in mathematics which is quite different from this [formalist] one. It means some statement in ordinary languge which has been proven to be true by a rigorous argument. ... But in

formal systems, theorems need not be thought of as statements — they are merely strings of symbols. And instead of
being *proven*, theorems are merely *produced*, as if by
machine.[125]

As Hofstadter readily admits, even to speak of *axioms* in the
context of a formal system marks a radical shift in the meaning
of the term.[126] Contrary to the guiding spirit of instrumentalist
apologies, however, far more than a matter of semantics is
involved here.

The 'theorems' in formal systems satisfy the syntactic
requirement that they are logical truths. Wittgenstein's purpose
in stressing that mathematical propositions are rules of grammar
*as opposed to* tautologies was to remove both elements of this
compositional picture.[127] Far from being the — cognitively
trivial — consequence of basic concepts together with the rules
governing their combination, we 'win through to a decision' in
mathematical proofs (RFM III §27) and use the resulting
mathematical propositions as norms for the transformation of
empirical statements: 'The proposition proved by means of the
proof serves as a rule — and so as a paradigm. For we *go by* the
rule' (RFM III §28). Much of *Remarks on the Foundations of
Mathematics* is devoted to clarifying the distinction between
logistic and mathematical theorems and *a fortiori* logistic and
mathematical proofs.[128] The argument harks back to 6.1263 of
the *Tractatus* where Wittgenstein distinguished between *proof
in logic* and *proof by logic*. The former does not demonstrate
that a logical proposition *follows* from other logical propositions
but is rather a procedure for producing logical propositions by
performing operations on the original symbols (TLP 6.126);
whereas the latter uses the rules of inference to establish the
truth of propositions. In his later work Wittgenstein remained
committed to the *Tractatus* account of 'Proof in logic [as]
merely a mechanical expedient to facilitate the recognition of
tautologies in complicated cases' (6.1262) but he came to hold
that a mathematical proof forges the internal relations of a rule:
'the proof changes the grammar of our language, changes our
concepts. It makes new connexions, and it creates the concept
of these connexions' (RFM III §31).

This normative conception of mathematics provides the core
of Wittgenstein's critique of Hilbert's Programme. Hilbert's
approach was guided by the assumption that formalization is a

rigorous extension of the processes originated by axiomatiz-ation. In fact it initiates a dramatic conceptual shift, and it is only by ignoring this transition that many of the problems embedded in Hilbert's conception of meta-mathematics can be inspired. At one and the same time we are asked to accept that the subject is intrinsically the same as mathematics even though its basic notions are *toto caelo* different. Far from protecting the theory from the embarrassing consequence that proofs, axioms, and theorems have been lost in the process, Hofstadter's inverted commas only advertise the fact that the truth of pure mathematics has been abandoned *ab initio* in Hilbert's Pro-gramme. Or at least, that the consistency of the axioms is abso-lutely crucial to the 'truth' of pure mathematics as conceived by Hilbert. Which brings us back to the ultimate focus in all this: the significance of Gödel's theorem *vis-à-vis* the certainty of mathematics as defined by Hilbert. And the very fact that the latter exposes us to a sceptical dilemma was sufficient proof for Wittgenstein of the need to re-examine the conceptual basis of Hilbert's Programme; in particular, Hilbert's fundamental premise that mathematical expressions can be divided into two classes: *real* (finitary) propositions whose epistemic utility is determined by their content, and *ideal* (transfinite) pseudo-propositions whose 'algebraic' properties facilitate our reason-ing about the former.[129]

In 'On the Infinite' Hilbert characterized the 'ideal pseudo-proposition $a + b = b + a$' as one which 'is no longer an imme-diate communication of something contentual at all, but a certain formal object'. It enables us to obtain 'finitary particular propositions' by substituting numerals for $a$ and $b$.

> Thus we arrive at the conception that $a$, $b$, $=$, and $+$, as well as the entire formula $a + b = b + a$, do not mean anything in themselves, any more than numerals do. But from that formula we can indeed derive others; to these we ascribe a meaning, by treating them as communications of finitary propositions. If we generalize this conception, mathematics becomes an inventory of formulas — first, formulas to which contentual communications of finitary propositions ... corre-spond and, second, further formulas that mean nothing in themselves and are the *ideal objects of our theory*.[130]

It is important to see that this argument remains committed to the referential picture of mathematical propositions. Indeed, the chief rationale for Hilbert's 'ideal pseudo-propositions' would appear to have been to avert Cantor's epistemological problems. For it was not infinity *per se* which troubled Hilbert, but solely the actual infinite.

This can be seen in his awkward distinction between generalizations and ideal pseudo-propositions. Unlike '$a + b = b + a$', '$[a] + [b] = [b] + [a]$ where $[a]$ and $[b]$ stand for specific numerals' is a real proposition which ranges over a potential infinity of numbers. But it is problematic in that its negation yields an ideal pseudo-proposition: one which 'cannot be interpreted as a combination, formed by means of "and", of infinitely many numerical equations, but only as a hypothetical judgement that comes to assert something when a numeral is given'.[131] The (constructive) reasoning behind this argument is clear enough, but however we interpret these 'generalizations' we are left with a problem: if they are treated as mathematical propositions we are faced with the impossibility of negating them; but if regarded as pseudo-propositions (whose instantiations yield meaningful propositions) we must then explain how they differ from ideal propositions. Conversely, there is the problem of explaining how a 'pseudo-proposition' expresses a 'hypothetical judgement'. (To paraphrase Wittgenstein's remark *à propos* Russell's 'complexes', Hilbert's 'ideal pseudo-propositions' had the useful property of being meaningless and combined with this the agreeable property that they could be treated like propositions.) Most perplexing of all is how *negation* could take an expression from one ontological realm into another, and in so doing cancel out its meaning.[132]

Wittgenstein's response to this argument was the antithesis of Frege's. Frege remonstrated that 'in the case of [Hilbert's] pseudo-axioms, there are no thoughts at all, and consequently no premises. Therefore when it appears that Mr. Hilbert nevertheless does use his axioms as premises of inferences and apparently bases proofs on them, these can be inferences and proofs in appearance only.'[133] But one hardly wants to deny mathematicians the use of the formulae characterized by Hilbert as 'ideal'. Unlike the instrumentalist attempt to circumvent 'Frege's Problem' meta-mathematically, however, Wittgenstein's strategy was to reconsider the latter conception. Rather than following Frege in the repudiation of Hilbert's 'ideal pseudo-propositions' he showed that they are nothing of the sort. To be sure, '$a + b = b + a$' does not operate in the same manner as '$2 + 3 = 3 + 2$'; but that hardly licenses the conclusion that it possesses *no* meaning. Nor does it substantiate Hilbert's faith in 'Cantor's paradise'.[134] For the difference here is normative, not epistemological; *both* propositions are rules of syntax: the former governing the use of '2' and '3' and the latter that of '+' and '='. Likewise, Hilbert's proposed distinction between real and ideal 'generalizations' collapses into the rules used to stipulate the basic concepts of a system, and it is entirely for reasons of logical grammar that 'we can indeed derive other' propositions from '$a + b = b + a$'. Hilbert was only led into his illicit categorial distinctions because he persisted in searching for ways to preserve the referential conception of mathematical propositions. Remove this framework and the need for a consistency proof does not occur because the issue of reliability does not arise.[135]

As was noted above, Hilbert saw the formalization argument, not as a means of removing intuition from mathematics, but rather of reducing higher mathematical truths to the haven provided by intuition. But 'of course, all of mathematics cannot be comprehended within this sort' of intuitively-grasped finitary proposition; 'the transition to the standpoint of higher arithmetic and algebra ... already denies these intuitive procedures.'[136] We cannot know that nonfinitary propositions are true independently of having a proof for them and we cannot have any faith in such a proof without a meta-mathematical proof which establishes that the former is sound. But the latter must itself be reliable, so the problem is to develop a method for reducing the justification of higher mathematical propositions to a format in which intuition can take over. Hilbert's claim was that such a step is made possible by the fact that a consistency proof is a meta-mathematical proof *about* the system in question, not a proof *within* that formal system. And since the formal system under meta-mathematical scrutiny is comprised of intuitively given symbols (when 'drained of all their meaning'[137]) the epistemological status of a consistency proof is the same as that of the self-evident propositions of elementary mathematics. In proceeding in this manner meta-mathematics would be created on exactly the same lines as governs mathematics proper: which is precisely the reason why Wittgenstein maintained that the problem with Hilbert's account of meta-mathematics lay in the very conception of *mathematics* which inspired his argument.

In *Philosophical Remarks* Wittgenstein insisted contra Hilbert that 'In mathematics, we cannot talk of systems in general, but only *within* systems. They are just what we can't talk about' (PR §152). The argument as presented sounds dogmatic, but it follows from the preceding clarification of the meaning of mathematical propositions as determined by intra-linguistic rules rather than a connection between language and reality. The point of this normative conception of mathematical propositions and proofs is to clarify that the meaning of a mathematical concept is not an object or 'configuration' but rather, the totality of rules governing the use of that concept in a calculus.[138] Mathematical propositions are not *about* anything (in a descriptive sense)[139] yet neither are they meaningless:[140] they are *norms of representation* whose essence is to fix the use of concepts in empiricial propositions.[141] Thus what most sharply separates Wittgenstein's conception of implicit definitions from Hilbert's (as expressed in the letter to Frege quoted above) is this emphasis that to be a mathematical proposition is to be applied. 'Meaningless strings', on the other hand, cannot be interpreted, only manipulated; all that Hilbert was entitled to conclude from his notion of mapping is that it is possible to construct rules for correlating meaningful statements with meaningless symbols. And that lends no credence to his assumption that mathematical notions can be applied *mutatis mutandis* to formal systems.

The transgression operating in Hilbert's argument was enshrined in his conception of axiomatics as the systematization of the body of knowledge contained in a theory, regardless of whether this be mathematical or scientific. One can only speak of 'systematizing the knowledge contained in a mathematical theory', however, by treating mathematical propositions as quasi-empirical, descriptive expressions: a convention which, despite his formalist leanings, Hilbert tacitly accepted. Hence Wittgenstein's conclusion that Hilbert had misconstrued the nature of meta-mathematics for the deeper reason that he misunderstood the very nature of mathematics. To argue that meta-mathematics constitutes a legitimate branch of mathematics whose 'objects of investigation' are proofs rather than numbers is to introduce yet another fresh confusion on the basis of the entrenched platonist assumption that mathematical propositions are descriptive. Yet that does not mean that meta-mathematical

211

proofs and propositions cannot be incorporated into the family of mathematics; only that if they are they cannot be *about* anything (in the referential sense), much less about mathematical proofs. Hence, the task remains to clarify the nature of the proofs constructed in this putatively new species of mathematics.

It was precisely this argument which led to Wittgenstein's blunt conclusion in his discussions with Waismann and Schlick that 'Hilbert's "metamathematics" must turn out to be mathematics in disguise' (WWK 136). For 'What Hilbert does is mathematics and not metamathematics. It is another calculus, just like any other one' (WWK 121). Hilbert's meta-mathematical proofs are not a higher, but simply a *different* calculus: 'I can play with the chessmen according to certain rules. But I can also invent a game in which I play with the rules themselves. The pieces in my game are now the rules of chess, and the rules of the game are, say, the laws of logic. *In that case I have yet another game and not a metagame*' (PR 319). It is for this reason that Wittgenstein objected to the central premise of Hilbert's Programme that philosphical problems could be solved (meta-)mathematically; for 'No calculus can decide a philo-sophical problem. A calculus cannot give us information about the foundations of mathematics' (PG 296). The key to Wittgen-stein's argument is thus the claim that Hilbert had merely constructed a new calculus which as such could in no sense be *about* the calculi of mathematics: it could not somehow *encompass* or *describe* an 'object calculus'.

That is not to preclude talking about formal 'theorems'. As Nagel and Newman point out: 'One may say that a "string" is pretty, or that it resembles another "string," or that one "string" appears to be made up of three others, and so on. Such state-ments are evidently meaningful and may convey important information about the formal system.'[142] But Nagel and Newman further suggest that such statements 'belong to what Hilbert called "meta mathematics," to the language that is *about* mathematics.'[143] If that were all that Hilbert meant by 'meta-mathematics' Wittgenstein would have had no objection. But none of the above examples qualify as *mathematical* propo-sitions: they are simply prose descriptions of 'formal strings'. In which case using the term 'meta-mathematics' for the 'description, discussion, and theorizing about [mathematical] systems'[144] would be extremely misleading. But, of course, that is not at all what Hilbert had in mind; for his Programme could only be successful by assuming at the outset that it is logically possible to step outside — and above — a system while staying within the broad confines of mathematics and achieve from this superior vantage point what had proved impossible from within the system. But if the referential conception of mathematical propositions which lies at the heart of this picture is removed we are left with a collection of calculi all of which stand on an equal footing. There is no hierarchy of calculi, just different — autonomous — calculi. Hence when we do revise or expand some existing system we are neither 'filling in a gap' nor ascending up some metaphysical ladder which directs our efforts to 'disclose' the body of the theory.[145]

Wittgenstein turned to the propositional and predicate calculi to illustrate his argument: 'The system of calculating with letters is a new calculus; but it does not relate to ordinary calculation with numbers as a metacalculus does to a calculus. *Calculation with letters is not a theory.* This is the essential point. In so far as the "theory" of chess studies the impossibility of certain positions it resembles algebra in its relation to calculation with numbers' (WWK 136). This latter remark was directed against the standard attempt to elucidate Hilbert's notion of meta-mathematics with the theory of chess.[146] Indeed, when he first heard Wittgenstein's argument Waismann objected:

There's a theory of chess, isn't there? So we can surely use this theory to help us obtain information about the possibilities of the game — e.g. whether in a particular position I can force mate in 8, and the like. If, now, there is a theory of the game of chess, then I don't see why there shouldn't also be a theory of the game of arithmetic and why we shouldn't apply the propositions of this theory to obtain material information about the possibilities of this game. This theory is Hilbert's meta-mathematics (PR 326).

We are now in a position to appreciate the full force of Wittgenstein's answer:

What is known as the 'theory of chess' isn't a theory describing something, it's a kind of geometry. It is of course in its turn a calculus and not a theory. .... [I]f in the theory I

213

use a symbolism instead of a chessboard and chess set, the demonstration that I can mate in 8 consists in my actually doing it in the symbolism, and so, by now doing with signs what I do with pieces on the board. When I make the moves and when I prove their possibility — then I have surely done *the same thing* over again in the proof. I've made the moves with symbols, that's all. ... [I]f I establish in the 'theory' that such and such possibilities are present, I am again moving about within the game, not within a metagame. Every *step* in the calculus corresponds to a move in the game, and the whole difference consists only in the physical movement of a piece of wood (PR 327).

It is obviously not a prerequisite of playing chess that one use a board with pieces; rather, it is that we make our moves in accordance with the totality of rules that constitutes playing chess. The mistake is to suppose that a 'symbolic description' of the game is in any way different from the actual moves in the game, as if the written expression 'K–Q3' were any different from moving the King one space on the board. Thus the 'meta-game' is either nothing more than *another* game (with different pieces and rules), or else it is the *same* game played with different signs standing for the same pieces.

When we turn to Hilbert's 'meta-mathematics' it emerges that the expressions in this putatively new species of mathematics must, *qua* mathematics, be rule-formulations fixing the use of these 'meta-mathematical' concepts. It is for this reason that Wittgenstein insisted that an expression like '"0 $\neq$ 0" is a theorem' is not a 'meta-mathematical' proposition about a mathematical proposition but rather, a rule-formulation expressed in the new calculus that Hilbert has constructed (prohibiting the deduction of '0 $\neq$ 0'):

If someone were to describe the introduction of irrational numbers by saying he had discovered that between the rational points on a line there were yet more points, we would reply: 'Of course you haven't discovered new points between the old ones: you have constructed new points. So you have a new calculus before you.' That's what we must say to Hilbert when he believes it to be a discovery that mathematics is consistent. In reality the situation is that Hilbert doesn't establish something, he lays it down. When

Hilbert says $0 \neq 0$ is not to occur as a provable formula, he
defines a calculus by permission and prohibition (PR 339).

If the status of meta-mathematics is confined to that of mathematics proper then the nature of meta-mathematical proofs —
whether limited to impossibility proofs or otherwise (for it is no
longer necessary to impose such arbitrary restrictions) — is confined to the construction of the rules which constitute a given
system (i.e. establishing what can and cannot be significantly
expressed in that system).

It would take us too far outside the scope of the present work
to see how Wittgenstein proposed to resolve/dissolve the consistency problem,[147] but we can at least appreciate how his
objection to Hilbert's Programme stemmed from his remarks on
the logical nature of mathematical propositions, and not from
any 'finitistic' misgivings. The point that Wittgenstein was
driving at is that Hilbert's notion of the need for a consistency
proof rested on his prior assumption about the proper method
for clarifying problems in the philosophy of mathematics (RFM
VII §16). In a passage which begins with the identical words
which he subsequently applied to Gödel's theorem Wittgenstein
warned:

If the contradictions in mathematics arise through an
unclarity, I can *never dispel this unclarity by a proof.* The
proof only proves what it proves. But it can't lift the fog.

This of itself shows that there can be no such thing as a
consistency proof (if we are thinking of the inconsistencies of
mathematics as being of the same sort as the inconsistencies
of set theory), that the proof can't begin to offer what we
want of it. If I'm unclear about the nature of mathematics, no
proof can help me. And if I'm clear about the nature of
mathematics, the question of consistency can't arise at all
(PR 320).

The point to focus on here is that such a philosophical objection
does not seek to challenge the technical cogency of Hilbert's
proof; Wittgenstein's sole concern was with the intelligibility of
the *problem* which Hilbert had embraced. His aim was to show
that Hilbert's Programme was the result of a sceptical muddle
which had led Hilbert into what is ultimately a mathematical
variant of the metaphysical urge to transcend the bounds of

sense. But while such a critique may have no bearing on the mathematical status of Hilbert's proof it has enormous significance in so far as the *interpretation* of that proof — and *a fortiori* Gödel's theorem — are concerned. It is time now to apply these considerations to Wittgenstein's reading of 'On Undecidable Propositions of Formal Mathematical Systems', bearing in mind throughout his intention to show that the 'proof only proves what it proves'.

## 6. GÖDEL'S PROOF

The first step in Gödel's proof — the assignment of Gödel numbers to the formulas and proofs of the system — is a straightforward method for establishing one–one correspondences between expressions in a calculus and an arbitrary set of signs (in this case, a subset of the integers; *infra*). There is no obvious reason for Wittgenstein to object to Gödel's method of arithmetizing a calculus, allowing him to determine for any given number whether it is a Gödel number and if so to calculate recursively from it to the corresponding expression in the calculus. For this clearly does not constitute meta-mathematics in the hierarchical sense. The Gödel numbers are not numerals which *denote* expressions in an 'object calculus'. What Gödel did was construct a new system (the GN-calculus of Gödel numbers) together with a set of rules for mapping expressions between the two calculi. But it is the next step in the argument — the 'arithmetization of meta-mathematics' — which brings in the first of Wittgenstein's objections. The premise operating here is that given the arithmetization of a calculus a meta-mathematical statement about the expressions in an object calculus can at the same time be read as a statement about the arithmetical relations holding between the corresponding Gödel numbers. In light of the arguments canvassed in the preceding section Wittgenstein regarded this premise as doubly misguided: to introduce meta-mathematical propositions at this stage must either be to reformulate the expressions of the original calculus in another notation, or else to construct another calculus to stand beside the two already in existence. In neither case can they be construed as propositions *about* 'object' expressions, and *a fortiori*, *about* the arithmetical relations between the corresponding expressions in the GN-calculus.

The importance of this point emerges when the role which this putatively minor premise plays in Gödel's interpretation of his proof is considered. Gödel wanted to show that meta-mathematical statements about the structural properties of the expressions within a calculus can be mapped and hence mirrored within those same expressions: given that a Gödel number can be assigned to every expression in the calculus a meta-mathematical statement about the logical relations between expressions in the calculus can be mirrored in the arithmetical relations between the expressions in the GN-calculus, which can be mapped onto the relations between the actual statements in the calculus. Hence by converse reasoning the arithmetical relations between the expressions in the GN-calculus can be interpreted as reflecting meta-mathematical propositions about the logical relations between these statements and meta-mathematical issues can be pursued by examining these arithmetical relations themselves. When Wittgenstein objected to this argument on the grounds that it is impossible to get outside from within a system because to do so would be to transcend the bounds of sense he was not criticizing the technical apparatus here established; he was demanding a careful reading of the candidates for such portentous meta-mathematical status.

Nagel and Newman offer as an example 'The sequence of formulas with Gödel number $x$ is a proof of the formula with Gödel number $z$'.[148] If this is seen as a contingently true (or false) proposition it must be one in which no grasp of the mathematical content of the proof/theorem is involved; that is, in which the relation between proof and proposition is external (as in e.g. the case of someone who points at a mass of figures on a board and announces: 'That is Gödel's proof of his famous theorem.') But if the statement is seen as necessarily — i.e. mathematically — true, then it can no longer be viewed in descriptive/empirical terms but must be treated as a rule of grammar (in this case stipulating the use of the terms 'proof' and 'theorem' vis-à-vis various figures in the GN-calculus). Gödel's interpretation — as opposed to his proof — moves subtly between these two alternatives, drawing on features of each type of expression. We are asked to accept that a meta-mathematical statement can be mirrored by a formula in the GN-calculus which depicts a purely arithmetical relation between $x$ and $z$. Gödel's meta-mathematical propositions are thus invariably treated as a species of — necessarily true — mathematical proposition; but like ordinary empirical propositions they are supposed to be *about* the logical relations holding in the 'object calculus' and are contingently true or false. To be sure, this exploits a feature which Hilbert built into his conception of meta-mathematics; but it is crucial to see how this ambiguity underpins Gödel's interpretation of his proof.

Wittgenstein cautioned in *Philosophical Remarks* that a 'useful question' in such situations is to ask 'How would this proposition actually be used in practice?' (PR §113). In this case he felt that a close reading reveals not descriptive propositions but, as expected, rules of logical grammar. When stripped of this spurious quasi-empirical gloss Gödel's 'meta-mathematical propositions' stand exposed as either prose reformulations of the rules of the 'object calculus' or rules that have been created for the coordination of expressions drawn from the 'object' and GN-calculus. The important point is that there are two distinct species of mathematical rules grouped together under the banner of meta-mathematics in Gödel's proof, each of which is called upon to perform a vital role in Gödel's subsequent interpretation. When the purportedly descriptive content of these propositions is called for the former type of disguised rules is pressed into service, and when the necessity of meta-mathematical truths is (tacitly) at stake the latter theme is emphasized.

The first type of use is typified by the M-sentence '¬P is the negation of P', where what is supposedly 'described' is a logical relation holding in the 'object calculus'. But the negation of such a statement yields not a false proposition but rather, an unintelligible expression. For this statement is not simply true: it is *necessarily* so. If the type-hierarchical assumption which implicitly treats this proposition as a statement about the logical relations obtaining in the 'object calculus' is removed what remains is a prose formulation of the rules governing the use of '¬'.[149] The second type of use is illustrated in Nagel and Newman's observation that 'a meta-mathematical statement which says that a certain sequence of formulas is a proof for a given formula is *true*, if, and only if, the Gödel number of the alleged proof stands to the Gödel number of the conclusion in the arithmetical relation here designated by ["Dem $(x, z)$"].'[150] But if that is the case their next sentence is seriously misleading, for they continue: 'Accordingly, to establish the truth or falsity of the meta-mathematical statement under discussion, we need concern ourselves only with the question whether the relation Dem holds between two numbers.'[151] But here too, given that one is dealing with manifestly necessary truths stipulating rules for mapping 'object' onto 'GN' expressions, then the contrary stipulation 'The sequence of formulas with Gödel numbers $x$ is not a proof of the formula with Gödel number $z$' renders the prior meta-mathematical expression *unintelligible*, not *false*.

This is not the sole problem with this argument, however, for in the process of stripping meta-mathematical propositions of their normative character a crucial shift has been introduced — on the basis of the two kinds of rules employed — from speaking of *mapping* to the concept of *mirroring* the content of one proposition in another. When dealing with rules for the co-ordination of 'object' with GN-calculus the notion of mapping obviously comes to the fore. But the notion of mirroring draws on (and demands) a tacit shift to the other type of rule — the prose reformulation of the rules of the 'object' calculus — without, it must be stressed, licensing the transition from *mapping* to *mirroring*. For these two notions are clearly not identical; whereas the former highlights the normativity of Gödel's meta-mathematical propositions the latter completely obscures it. The type-hierarchical premise operating here perforce assumes that the referential content of a 'higher' proposition can actually be read — given a mastery of the rules of projection — from the structural properties of the 'lower' propositions. Nagel and Newman assume that

> The reader will see readily that the expression 'sub (y,13,y)' is the mirror image *within* the formalized arithmetical calculus of the meta-mathematical characterization: 'the Gödel number of the formula that is obtained from the formula with Gödel number $y$, by substituting for the variable with Gödel number 13 the numeral for $y$'.[152]

Yet such a conclusion is by no means licensed by the preceding argument.

Given a mastery of the meta-mathematical rules what the reader will readily see is that 'sub (y,13,y)' must be *mapped* onto the appropriate 'object' proposition in light of the meta-mathematical rule partially described. To pass beyond this into the notion of *mirroring* is but another manifestation of the assumption that meta-mathematical propositions constitute a logical hybrid of mathematical and factual statements. Whereas 'P is the negation of $\neg$P' is neither the *mirror* image of nor *mapped* onto

| $P$ | $\neg P$ |
|---|---|
| T | F |
| F | T |

but rather, is the same rule, stipulated in prose rather than by truth-table. Admittedly these are entirely problems of *interpretation* rather than construction; but that in itself is an extremely serious matter. Some idea of the force of this charge comes out in Gödel's preface in which he maintained that the proof establishes that 'From the remark that [R(q);q] says about itself that it is not provable it follows at once that [R(q);q] is true, for [R(q);q] *is* indeed unprovable (being undecidable). Thus, the proposition that is undecidable *in the system PM* still was decided by meta-mathematical considerations.'[153] It is all too easy to overlook the radical philosophical claim here presented. Contained in these brief words is not simply an obituary of Hilbert's Programme (and with it a covert plea for platonism); at stake is our understanding of the logical character of mathematical propositions: of the nature of the relation between a mathematical proposition and its proof.

Turning from prose to proof it appears that the key to this striking result lies in the initial construction of an arithmetical formula P that 'represents' the meta-mathematical statement 'P is unprovable'. That is, the proposition [R(q);q] ('P') represents *within* the calculus the meta-mathematical proposition 'P is unprovable'. (Or as Wittgenstein insisted on describing it, following Gödel's precedent, 'P is not provable in Russell's system'.) In Gödel's prose 'We therefore have before us a proposition that says about itself that it is not provable.' Interestingly Nagel and Newman are more circumspect in their account of this point; they explain that 'the *arithmetical formula* "(x)$\neg$Dem (x,sub (n,13,n))" *represents* in the calculus the *meta-mathematical statement*: "The formula '(x)$\neg$Dem(x,Sub (n,13,n))' is not demonstrable." In a sense, therefore, this arithmetical formula G can be construed as asserting of itself that it is not demonstrable.'[154] But this odd sense of self-assertion — and Nagel and Newman's visible discomfort — is due to the fact that it is based on the shift from *mapping* to *mirroring*, which has now been taken a step further into the metaphysical realm of what Hofstadter describes as 'strange loops' (where 'sense' is defined as an infinitely ascending hierarchy of languages, the upper reaches of which are accessible from the lowest of linguistic levels).

In the next step of the proof Gödel set out to establish that P is provable if and only if ¬P is, and thence, that if ¬P is provable arithmetic is ω-consistent. But it cannot simply be assumed at this stage that P can be treated as a significant albeit unprovable proposition; a fallacy to which one is particularly prone when working within the context of 'formal systems'. All that the argument has so far established is that it is possible to construct well-formed strings of symbols according to the rules of syntax (itself a dubious notion as far as propositional status is concerned) which are not 'derivable' according to the transformation rules of the system. But that, according to the arguments pursued in the preceding section, does not warrant speaking here of a *proposition*.[155] It is the step immediately following on this, however, which creates the major philosophical problem posed by his proof, for Gödel set out to show that although it is not provable P is nonetheless true: i.e. even if it is undecidable P is nonetheless a genuine mathematical proposition. Should this step be successful then the relation between a mathematical proposition and its proof must indeed be purely external: a crucial philosophical result that will have been established by strictly mathematical means.

Given the themes so far examined it should come as no surprise to discover that the key to this startling development lies in the notion of meta-mathematics as conceived by Hilbert. For the crux of Gödel's argument is that assuming that arithmetic is consistent the meta-mathematical statement 'P is unprovable' is true. Given that this statement is merely the meta-mathematical reflection of 'P' itself within the calculus, if 'P is unprovable' is true this can only be because it has been mapped onto the true arithmetical proposition 'P'. In short, if 'P' has been mapped onto a true meta-mathematical statement it *must* be true. But this is precisely the step which Wittgenstein disputed, not on the basis that what is questionable here is the truth or falsity of P but rather, that 'P' is in fact *intelligible*. Far from committing a mathematical oversight Wittgenstein was arguing that we cannot draw any philsophical consequences from Gödel's formal (two-tier) version of the *pseudomenos* because it is incoherent to assert that 'P is unprovable and true'. What we have to clarify, therefore, is the philosophical as opposed to mathematical manner in which Wittgenstein intended his argument to be read.

Since Wittgenstein's objection strikes at this stage of Gödel's proof there is little reason to examine the final steps in Gödel's proof (the conclusion that since P is unprovable and true PM is essentially incomplete and that the consistency of arithmetic cannot be established by an argument that can be represented in the formal arithmetical calculus, although this does not preclude non-finitary meta-mathematical consistency proofs). For if Wittgenstein's critique is successful neither of these last two steps goes through. Thus we can now see why those who have argued that Wittgenstein unknowingly sacrificed an important potential ally in Gödel have missed the point of Wittgenstein's attack. Their feeling is that because of his assault on Hilbert's Programme and the emphasis on surveyability which he placed in his account of proof Wittgenstein should have welcomed Gödel's demonstration that it is impossible to give a finitistic meta-mathematical proof of consistency. What they fail to recognize is that, for Wittgenstein, the dilemma posed by Gödel's result is specious because the issue addressed by Gödel is spurious: Gödel's *interpretation* is thus vitiated by the fact that he adopted the framework of this illicit problem.

It is frequently claimed that Wittgenstein was guilty of confusing the levels of discourse operating in Gödel's proof. Gödel distinguished between the actual sentences of the formal system under examination (P-sentences), the arithmetical statements that express the content of P-sentences (A-sentences), the numbers associated with P-sentences (GNs), the meta-mathematical statements about these P-sentences (M-sentences), and finally, the correlation of A-sentences with M-sentences. For example, suppose that $P_1$ has GN i, and the negation of $P_1$ ($\neg P_1$) has GN j; then the M-sentence '$\neg P_1$ is the negation of $P_1$' is correlated with the arithmetical statement 'jRi'. The supposition is that Wittgenstein misunderstood Gödel's theorem because Gödel did not take any great pains to make the various levels of statement that the proof employs perspicuous. (Gödel did not actually write down P-sentences but referred to them by their GNs, which themselves are not written out but are denoted by the complex functions for their calculation which are also used to denote the M-sentences to which they correspond.) It is this criticism of Wittgenstein's remarks which, perhaps, demands closest scrutiny.

Without question one of the most intriguing aspects of Gödel's proof is the manner in which he sought to demonstrate that the recursive arithmetical statements (A-sentences) which are correlated with M-sentences can themselves be expressed within the P-sentences of the formal system P and thus, that to the *truth* of A-sentences corresponds the *provability* of the corresponding P-sentence (in so far as these complex recursive functions are built up out of simpler functions which themselves are all expressible in P). That is, in terms of Gödel's undecidable sentence (U), $A_i$ states that there is no factorial successor of i. If that were false then i would have a factorial successor; i.e. it would be the case that the P-sentence with $GN_i$ (U) could be proved in P, which has been ruled out by $M_i$. Therefore, U is undecidable while $A_i$ is true; U expresses within P the arithmetical statement $A_i$ which in turn is correlated to the meta-mathematical statement $M_i$ which states that U is unprovable. And this, it is urged, is the argument which Wittgenstein misunderstood.[156] Wittgenstein, so the charge goes, failed to recognize the reason why although Gödel's proof is comparable to it is not identical with the *pseudomenos* or the *Epimenides*. Whereas the latter both generate a paradox because they are self-referential Gödel's paradoxical result rests on a stratified proof in which none of the three key statements is self-referential:

U: '(x)(The number x is not factorially succeeded by the number i)'

$A_i$: 'There is no factorial successor to the number i'

$M_i$: 'The P-sentence with $GN_i$ (U) is unprovable'

Hence unlike traditional semantic paradoxes U is constructed and acquires its significance in the same way as any other P-sentence, without referring to the M-sentence with which it is correlated. The predicate of U is correlated with the meta-mathematical concept of *provability*, but U is undecidable independently and quite regardless of this arithmetization; even if there were no meta-mathematical system M, U would still be undecidable. Thus U is not *explicitly* self-referential: although U has the $GN_i$, it does not *state* that it does.

Is Wittgenstein's criticism vitiated by the possibility that he did not perceive the distinctions Gödel drew between the levels of discourse in which U, $A_i$, and $M_i$ occur? Certainly the discussion in RFM I APP III makes no rigorous attempt to follow through the intricate steps of Gödel's proof, thereby oversimplifying well beyond the point of opacity. But someone who concluded from this that Wittgenstein failed to understand the mechanics of Gödel's proof — or worse, that he advanced a 'completely trivial and uninteresting misinterpretation' of Gödel's results[157]) — could only do so at the cost of neglecting the extensive discussions of the logical character of mathematical propositions and proofs which provide the framework for Wittgenstein's critical remarks on Gödel's interpretation of his theorem. As we have seen, Wittgenstein deliberately rejected any such hierarchical distinction: before he is found guilty of technical deficiency we must examine how this theme applies in this respect, mindful of the fact that Wittgenstein's remarks on Gödel's theorem only make sense within the context of his earlier attack on Hilbert's notion of meta-mathematics and not vice versa. For as was stressed above, Wittgenstein removed the meta-mathematical focus of his early critique of Hilbert's notion of an absolute consistency proof and substituted for it the topic of greatest current interest — Gödel's incompleteness results — while leaving the substance of his original argument intact.

Wittgenstein's response to these developments was to cut them off before they can get started. The crux of his argument is that the illusory appearance of Gödel's paradoxical prose conclusion that a mathematical proposition can intelligibly say of itself both that it is unprovable *and* that it is true stems from the premise that one can distinguish between U and $M_i$ and treat them as autonomous mathematical propositions mapped onto one another, and then argue further that the structures of U and $M_i$ are such that they mirror one another. In seizing on Gödel's theorem as an illustration of the confusions which result from the belief that there are 'gaps' in mathematics Wittgenstein demonstrated his grasp of a point which few others have fully appreciated. Far from legitimizing Hilbert's conception of the mathematics/meta-mathematics distinction, Gödel's interpretation of his proof rests on the premise that Hilbert's demarcation is self-evident. But this was no oversight or casual assumption on Gödel's part. For he intended to accomplish more than prove the existence of formally undecidable propositions. As John Dawson has documented, Gödel was aware of the *negative* impact of his theorem on Hilbert's Programme, but even more important to him was the *positive* role which he hoped his proof would perform: if genuinely successful it would utilize 'Hilbert's meta-mathematical progeny' to revitalize platonism in the midst of the positivist mood dominating analytic philosophy in the 1930s (cf. Chapter 4 above).

In the companion paper published shortly after his proof Gödel drew attention to the epistemological framework on which his argument rested:

the undecidable propositions constructed for the proof of Theorem 1 become decidable by the adjunction of higher types and the corresponding axioms; however, in the higher systems we can construct other undecidable propositions by the same procedure, and so forth. To be sure, all the propositions thus constructed are expressible in Z ... [T]hey are, however, not decidable in Z, but only in higher systems.[158]

This is an explicit statement of the central premise underpinning Gödel's interpretation of his theorem: not only is it possible to prove undecidable propositions in higher systems, but equally significant, any of the undecidable propositions that can be constructed in higher systems are already expressible at the lowest level. This brief addendum is no mere exegesis; nor is it an apology. Gödel's purpose in 'On Completeness and Consistency' was to ensure that the hierarchical/meta-mathematical implications of his work as he perceived them had been grasped: not just in regard to the completeness of formal systems for arithmetic, but even more importantly, in highlighting the scope of the meta-mathematical domains waiting to be explored by mathematical logicians.

The reaction to his theorem must have more than satisfied Gödel's expectations. Ivor Grattan-Guinness has drawn attention to the fact that 'While Gödel's theorem rebuffed Hilbert's hopes, it did not detract from interest in meta-mathematics; indeed it led to interest in non-finitary consistency proofs.'[159] But one can go further; in Grattan-Guinness's own words: 'The contributions of Gentzen and Herbrand exercised a marked influence on the development of post-Gödelian meta-mathematics. Indeed, meta-mathematics became one of the chief interests of logicians in the inter-war period.'[160] Then there is the pivotal role which Gödel's proof played in the evolution of recursion theory. For Gödel had discovered that all general recursive functions are calculable in the formal system P, and can all be given explicit arithmetical definitions (e.g. using only addition, multiplication, and elementary logic with identity). Apart from — or perhaps because of — the problem that the notion of a 'general recursive function' was disturbingly vague, Gödel's proof stimulated a widespread interest in the power of the theory to provide recursive functional formulations for all number-theoretical problems. Finally, Gödel's theorem has indeed served as the vehicle for a marked revival of platonism. It thus takes on a pivotal role in the evolution of mathematical logic comparable to any of the transitional impossibility proofs considered in the preceding sections. It is hardly surprising, therefore, given the philosophical dimensions of these developments, that Wittgenstein should have seized on Gödel's theorem as the focus of his anti-metaphysical critique;[161] nor that his strategy should have been to strike at the hierarchical/meta-mathematical premise which he saw as the foundation of Gödel's argument. For Gödel's conception of mathematical discovery — as expressed for example in the above passage — articulates the very principle which Wittgenstein had sought to undermine in *Philosophical Remarks*. Gödel's picture is one of constantly expanding systems which enable the mathematician to fill in the gaps that have rendered certain propositions undecidable in lower systems.[162] Hence his argument represents the epitome if not the apex of the major themes which Wittgenstein challenged in the philosophy of mathematics.

Unfortunately, Wittgenstein failed to devote the sort of attention to his attack which would have been necessary if it was to gain the serious attention of mathematical logicians. Moreover, if — as is invariably the case — they are read out of context. Wittgenstein's remarks are baffling (to say the least). Finally, the consequences of Wittgenstein's argument are far more complex than is suggested by his brief discussion. Unlike the isolated criticisms of Gödel's theorem which surfaced in the 1930s (*infra*) Wittgenstein did not seek to identify any fault in the mathematical reasoning of the proof, nor does his objection limit us to a single alternative reading of Gödel's theorem. For there are radically different directions in which his argument can develop, depending on which species of meta-mathematical

226

rule is at stake. Wittgenstein did not offer any guidelines, however, as to how Gödel's theorem *should* be read, nor did he venture any comments on where the mathematical significance of Gödel's theorem, when properly understood, might lie. His sole task as he conceived it was to clarify the nature of the framework from which any of the various possible interpretations of Gödel's theorem must proceed if they are to remain intelligible. After that it was up to mathematical logicians to ascertain the significance of Gödel's proof or the uses to which it might be put.

In order to redress the damaging impression left by Wittgenstein's highly schematic critique the issue that must be addressed is how the various themes examined above apply in an actual reading of Gödel's proof. Given that it makes no sense to speak of a meta-mathematical proposition *about* an 'object mathematical proposition' what remains to be clarified is how Gödel's theorem should be interpreted when (i) what we are dealing with are autonomous mathematical systems, and (ii) we are simply confronted with prose reformulations of an existing system. The former example is perhaps the more subtle of the two. In so far as the two systems are autonomous there is no problem in treating U and $M_i$ as *parallel* mathematical propositions; but can we go further? Everything here turns on the notion of mirroring which rests on the prior assumption that $M_i$ is about U. The problem thus comes down to the question of what remains once this premise is removed. According to Gödel there are two independent calculi — P and M —, two independent propositions — U and $M_i$ — the former of which is unprovable while the latter is true, and a set of rules which have mapped U onto $M_i$. Of course, the rules may have been so constructed that they only map true propositions from each system onto one another, but *that* is no reason to conclude that U is true but unprovable; on the contrary it simply entails that the rules have broken down in this instance. One can only get to Gödel's much stronger conclusion by shifting from the notion of mapping to mirroring. Without this one is merely left with two independent propositions neither of which can meaningfully be expressed in the other's system.

As was touched on parenthetically above, Wittgenstein seized on Gödel's emphasis that 'the proposition that is undecidable *in the system PM* still was decided by meta-mathematical considerations'. Wittgenstein saw in this a reflection of the central theme in his proposed resolution of the decision problem:

> Weyl puts the problem of decidability in the following way. Can every *relevant* question be decided by means of logical inference? The problem must not be put in this way. Everything depends on the word 'relevant'. For Weyl, a statement is relevant when it is constructed from basic formulae with the help of seven principles of combination. ... This is where the mistake lies. A statement is relevant if it belongs to a *certain system*. It is in this sense that it has been maintained that every relevant question is decidable. What is not visibly relevant, is not relevant at all (WWK 37).

This is not an epistemological thesis, but rather a clarification of the logical grammar of *mathematical proposition*: of the rules which entail that to be a significant mathematical proposition is *ipso facto* to be decidable. Thus Wittgenstein asked: 'Does the question of relevance make sense? If it does, it must always be possible to say whether the axioms are relevant to this proposition or not, and in that case this question must always be decidable, and so a question of the *first* type has already been decided. And if it can't be decided, it's completely senseless' (PR §149). Whether one is dealing with the decision problem or Gödel's proof, therefore, the same question should be asked in each case: what does it mean to say that a *proposition* is true but unprovable *in this system*? What sort of 'proposition' could this be?

Gödel's argument pushes us to accept that there are two versions of the *same proposition* which is true but unprovable in one system while true and provable in another. But the whole point of Wittgenstein's argument on the autonomy of mathematical systems is that a mathematical proposition is *internally* tied to its proof/proof-system: 'If, then, we ask ... "Under what circumstances is a proposition asserted in Russell's game?", the answer is: at the end of one of his proofs, or as a "fundamental law" (Pp.). There is no other way in this system of employing asserted propositions in Russell's symbolism' (RFM I APP III §6). To be a mathematical proposition just is to be a member of a given system; mathematical propositions cannot be transposed from one system to another for the rules formulating the primitive concepts may differ. To be sure, the same sentence can be used to express manifest or — as in the case of over-lapping systems — subtle shifts in meaning; the crucial point to observe is the grammatical autonomy of such sentences. Thus Wittgenstein asked:

> 'But may there not be true propositions which are written in this symbolism, but are not provable in Russell's system?' — 'True propositions', hence propositions which are true in *another* system, i.e. can rightly be asserted in another game. Certainly; why should there not be such propositions; or rather: why should not propositions — of physics, e.g. — be written in Russell's symbolism. The question is quite anal-ogous to: Can there be true propositions in the language of Euclid, which are not provable in his system, but are true? — Why, there are even propositions which are provable in Euclid's system, but are *false* in another system. May not triangles be — in another system — similar ( *very* similar) which do not have equal angles? — 'But that's just a joke! For in that case they are not "similar" to one another in the same sense!' — Of course not; and a proposition which can-not be proved in Russell's system is 'true' or 'false' in a dif-ferent sense from a proposition of *Principia Mathematica*. (RFM I APP III §7)

In short, the meaning of a mathematical proposition is strictly determined by the rules governing its use in a specific system. If dealing with autonomous calculi then no matter how similar the rules of the two systems might be, as long as they differ — as long as we are dealing with distinct mathematical systems — it makes no sense to speak of the *same* proposition occurring in each. The most that can be concluded is that *parallel* propo-sitions occur in the two systems which can easily be mapped onto one another.[163] Hence Gödel was barred by virtue of the logical grammar of *mathematical proposition* from claiming that he had constructed identical versions of the same mathematical proposition in two different systems. The most he could justi-fiably have concluded is that he had constructed parallel mathe-matical propositions. But in that case there is nothing wrong with discovering that one member of this pair is unprovable in its system while the other is true. Indeed, as was seen in the early sections on impossibility proofs, the whole point of con-structing a new system is frequently exactly this: to complement an existing calculus with some new rule which enables us to transform a meaningless expression into a significant — i.e. decidable — mathematical proposition (a phrase, it should be noted, which is strictly redundant). In which case Gödel's proof can simply be read as an example of the internal pressures which drive a mathematician to expand an existing calculus — thus creating a new calculus in its place — in order to prove a proposition such as U.

This is, perhaps, the less interesting of the two cases, for it will naturally be objected that even if such a conclusion did arise *vis-à-vis* two autonomous calculi the important matter is, as Gödel showed, when there are two versions of the same proposition. Here too the problem must be approached without descriptivist presuppositions; but it is complicated by the question whether this is meant to be a mathematical or prose proposition. Wittgenstein scrutinized the M-sentence 'The P-sentence with $GN_i$ is unprovable': again in light of the critique of Hilbert's conception of meta-mathematics. On Gödel's interpretation $M_i$ states: 'U is unprovable'; i.e. '"$(\forall x) \neg (xFSi)$" is unprovable'. The brunt of Wittgenstein's argument is that to describe a mathematical expression as unprovable is to deny that it is a *mathematical proposition*: i.e. that it is intelligible. In which case '"$(\forall x) \neg (xFSi)$' is strictly comparable to an expression such as '$(\forall n)(x^n + y^n = z^n)$': in the absence of a proof this is not a meaningful yet unproved *proposition* but rather, a meaningless expression albeit one which may exercise a heuristic influence on the construction of some new proof-system. The statement '"$(\forall x) \neg (xFSi)$" is unprovable' is not a meta-mathematical proposition about U but rather, a prose formulation which, as with the statement '"Fermat's last theorem" is unprovable', asserts that a given mathematical string is unintelligible. It is only the illicit assumption that the content of an 'object' can be *mirrored* in a 'higher' meta-mathematical proposition which inspires the belief that the relation between a mathematical proposition and its proof is external: and thence the epistemological discussions of the significance of Gödel's theorem.

If $M_i$ is seen as a proposition it must operate in the same system as U. Here Wittgenstein's point is that Gödel's theorem is more than just similar to such contradictions as the *pseudomenos* or the *Epimenides*; once Hilbert's framework is removed it is apparent that, if treated as a genuine mathematical as opposed to prose assertion $M_i$ must indeed be self-referential. For 'Let us suppose I prove the unprovability (in Russell's system) of P; then by this proof I have proved P. Now if this proof were one in Russell's system — I should in that case have proved at once that it belonged and did not belong to Russell's system' (RFM I APP III §11). That is, if $M_i$ is to be interpreted as a mathematical proposition it must be one which reduces to a straightforward contradiction of the form: '$(\forall x) \neg (xFSi)$ & $(Ex) (xFSi)$'. U expresses an undecidable — i.e. meaningless — expression, and $A_i$, which is true if and only if $M_i$ is true is likewise undecidable; for $M_i$ *qua* mathematical proposition is neither true nor false but one in which all information is cancelled out. And this, ultimately, is the reason why Wittgenstein concluded that Gödel's proof shows (to paraphrase Wittgenstein) that 'There is no factorial successor to the number i, so there is a factorial successor to the number i, so there is no factorial successor to the number i, ... It is a profitless performance! — It is a language-game with some similarity to the game of thumb-catching' (RFM I APP III §12). That is not to say that Wittgenstein dismissed Gödel's theorem, however; only, that he attempted to alter our interpretation of its significance: the final point of his discussion to which we can now turn.

## 7. WITTGENSTEIN'S SERVICE TO THE PHILOSOPHY OF MATHEMATICS

It is helpful to consider, when assessing the import of Wittgenstein's remarks on the significance of the first incompleteness theorem, how they relate to some of the debates which Gödel's proof has aroused. There is an obvious affinity between one line of the preceding argument and the objection advanced by Perelman in 1936 that Gödel had only succeeded in constructing 'a new antinomy with identical structure to the well-known classical paradoxes ... as a result of a contradiction posed in the premises'.[164] Dawson would find it 'tempting to dismiss Perelman as a crank, were it not that his arguments were apparently taken quite seriously by many within the mathematical community'.[165] For Perelman based his attack on Gödel's 'informal introductory arguments' rather than the 'formal proof' which brought about 'a rather obvious conflation of object- and metalanguage'.[166] But if the brunt of Wittgenstein's critique is sound Perelman was in a sense quite correct to proceed in this manner, in so far as the *philosophical* substance of Gödel's argument is contained in his preface and not the body of the proof. Nevertheless, if Perelman did arrive at an aspect of Wittgenstein's argument it was for entirely the wrong reasons; for Perelman appears to have been oblivious of the framework of Gödel's proof, whereas Wittgenstein was concerned with little else.[167]

Perelman might safely be ignored,[168] but Zermelo's similar conclusion that Gödel's proof merely leads 'to a *contradiction* analogous to Russel[l]'s antinomy' cannot so easily be dismissed.[169] Zermelo approached Gödel's theorem from a completely different perspective than Perelman; where the former's interest was in a type-hierarchical elimination of the paradoxes,[170] Zermelo sought to salvage his infinitistic conception of logic which demanded that every true proposition be provable on its appropriate level of quantification. Perelman may have failed to recognize the significance of the object/meta-mathematics distinction but no such charge can be levelled against Zermelo.[171] Certainly Gödel did not press it; the main theme of their 1931 exchange devolved onto Gödel's finitistic notion of proof. Gödel did not dispute Zermelo's claim that undecidable propositions may always be decidable in higher systems; on the contrary, he stressed this himself in 'On Completeness and Consistency' (which may in part have been written in order to forestall Zermelo's objection). Gödel emphasized 'that for me the essential point of my result is not that one can somehow or other exceed every formal system (which already follows from the diagonal procedure) but rather that for every formal system there is a mathematical proposition which can be *expressed* within the system but *cannot be decided* from the axioms of the system.'[172] But Zermelo persisted that 'all that you have proved in your paper comes down to this, which I have always stressed, that a "finitistically limited" proof-schema does not suffice to "decide" every proposition of an uncountable mathematical system.'[173] In other words, that Gödel's 'undecidable proposition' was decidable precisely because it was *not* expressible in Gödel's formal system.

The most notable feature of Zermelo's criticism — as far as the present discussion of Wittgenstein is concerned — is how closely it accords with the theme just examined. Admittedly Zermelo's conception of mathematical propositions — which follows from the type-hierarchy on which Zermelo-Fraenkel set theory is based — differed radically from Wittgenstein's. But apart from this fundamental divergence Zermelo's criticism came down to the similar point that, as he wrote to Gödel on 21 September 1931:

> your proof of the existence of undecidable propositions exhibits an essential gap. In order to produce an 'undecidable' proposition, you define ... a 'class sign' (a propositional function of *one* free variable) S = R(q), and then you show that neither [R(q);q] = A nor its negation ¬A would be 'provable'. But does
>
> $$S \equiv \overline{Bew}[R(n);n]$$
>
> really belong to your 'system'?

In his response Gödel concentrated on the significance of the meta-mathematical framework underlying his theorem. But if what is denied is not the finitistic limitation under which Gödel's proof operates but rather, the Hilbertian premise which underpins Gödel's interpretation, then this is precisely the criticism which Wittgenstein raised. And unlike Perelman and Zermelo Wittgenstein's 'conflation' of object- and meta-language was completely deliberate, and even more important, Gödel's 'informal introductory arguments' have quite pronouncedly displaced the formal proof as the focus of *philosophical* investigation.

Perhaps the most difficult aspect of this proposed shift to reconcile with the interests of mathematical logic is how such dramatic consequences could be reached on the basis of a critique that proceeds long before the realm of Gödel's proof is entered. Indeed, none of the foregoing has any impact on the cogency of Gödel's proof; all this is primarily concerned with the first two pages of prose prefaced to the proof. Shorn of this philosophical overture a *bona fide* proof still remains; but it may well turn out to be of little more interest than any other paradox or, on the alternative reading, a closure impossibility proof which like all members of the genre must be judged solely on the importance of what it prohibits (RFM I APP III §14). This last point constitutes the crux of Wittgenstein's philosophical quarrel with Gödel. While it removes the metaphysical over-tones of the incompleteness theorems Wittgenstein's argu-ment leaves intact the widespread conviction that Gödel succeeded in undermining Hilbert's Programme; but it demands a significantly altered attitude to the latter accomplishment. On Gödel's interpretation the first incompleteness theorem establishes that the relation between a mathematical proposition and its proof is external, thereby transforming his results into a transitional impossibility proof. Wittgenstein's reading confines it to the role of closure impossibility proof. By emphasizing the purely logical point that this relation is *internal* — that a mathe-matical proposition is *inextricably* tied to its proof — Gödel's theorem serves as a *reductio ad absurdum* of the philosophical framework underpinning Hilbert's Programme: viz. by forcing us to accept that a mathematical theorem could *per impossibile* be intelligible *prior to* or *independently of* the construction of the set of rules which create its meaning.

When seen in this light Wittgenstein's argument does not in the least belittle the genius whereby Gödel achieved this feat. Here is a case in point of the subtlety with which Wittgenstein felt philosophy must draw its conclusions. Often all that is called for is a modest change in point of view; alter the focus slightly and Gödel's proof will not have lost any of its significance: only the platonism will have been abandoned. Which recalls the sug-gestion that Wittgenstein should have recognized a potential ally in Gödel. In a sense this is correct, but in very much the oppo-site direction from the manner intended. For Wittgenstein was in no need of an epistemological comrade-in-arms; on the con-trary he was striving to undermine what he saw as the spurious premises inspiring the foundations crisis. One might thus argue that if 'On Formally Undecidable Propositions' is purged of Hilbert's epistemological bias a formidable mathematical ana-logue of Wittgenstein's philosophical critique of Hilbert's Pro-gramme emerges. Yet to do so would be to run contrary to the spirit in which Gödel regarded his results. For as Dawson and Feferman have shown, Gödel harboured platonist convictions long before the discovery of his incompleteness theorem: which is all the more reason to clarify the philosophical overtones of his proof as contained in its prose. To accomplish this, however, ultimately leads one outside the bounds of Gödel's proof into his subsequent work on the Continuum Hypothesis, where he first explicitly formulated what he saw as the platonist implications of the incompleteness theorems.[174] To be fair to Gödel, however, whatever epistemological theses he developed in later life his original interpretation of his proof was no less restricted to the logical grammar of *mathematical proposition* than Wittgenstein's response. And unlike Hardy, who resorted to specious epistemology in order to posit the existence of transcendental truths, Gödel relied on logic itself to demonstrate their presence; the epistemology, important as it was, only came later.

There is an understandable tendency to suppose that by announcing his desire to by-pass Gödel's proof Wittgenstein was indicating the low esteem in which he held Gödel's work. Yet had that been the case Wittgenstein would quite literally have by-passed Gödel's theorem: viz. by ignoring it. But as was remarked at the outset of this paper, how could anyone let alone someone as intent on scrutinizing the philosophical significance of mathematical logic as Wittgenstein seriously hope to avoid one of its foremost monuments? Certainly it was not Wittgenstein's intention, and the very manner in which he discussed Gödel's theorem reflects the seriousness with which he treated Gödel's results. To by-pass in this case, therefore, means to escape the consequences which Gödel's theorem seemingly thrusts upon us; which in itself is proof of the overriding importance of Gödel's theorem to the philosophy of mathematics: indeed to philosophy in general. It would thus be a travesty of Wittgenstein's approach — similar to that committed by Russell in his response to the *Investigations* — to suppose that because he challenged Gödel's interpretation Wittgenstein in any way intended to 'trivialize' Gödel's proof. For to dethrone Gödel's theorem is by no means to demean it; only to divest it of its false metaphysical pretensions.

On a variant reading *Remarks on the Foundation of Mathematics*

is haunted by the specter of Gödel; it is a dramatic presentation of the writhing struggle in which Wittgenstein as logical Laocoön tries to escape annihilation by the Gödelian serpent. Less histrionically stated, it provides a living demonstration of how Wittgenstein, through a reinterpretation of the nature of mathematics made necessary by Gödel, is forced before our very eyes to pass from the 'logistic' to the 'linguistic' point of view.[175]

But this too completely misses the point of Wittgenstein's remarks: the framework of which, as was stressed above, was already in place before the announcement of Gödel's theorem. Rather, Wittgenstein's 'struggle' was to show how the nature of mathematics forces a reinterpretation of Gödel's theorem (RFM VII §22).

A somewhat more troubling matter is the absence of any mention in *Remarks on the Foundations of Mathematics* of the possibility of deducing Gödel's incompleteness theorems from other results; e.g. consistency proofs of primitive recursive arithmetic. This is particularly puzzling when seen in the light of R.L. Goodstein's account of

the mystery that what Wittgenstein said on the subject [of Gödel's theorem] in 1935 was far in advance of his standpoint three years later. For Wittgenstein with remarkable insight said in the early thirties that Gödel's results showed that the notion of a finite cardinal could not be expressed in an axiomatic system and that formal number variables must necessarily take values other than natural numbers; a view which, following Skolem's 1934 publication, of which Wittgenstein was unaware, is now generally accepted.[176]

It is not just the 'principle of charity' which suggests that, far from being a lapse, this putative lacuna reflects Wittgenstein's conscious decision to concentrate on the issue which he felt was of utmost *philosophical* importance: the logical character of mathematical propositions. The problem here is not so much how one arrives at Gödel's theorem as how such a theorem should be interpreted (RFM I App III §17). But that is not to say that the background for which Goodstein looked in vain is indeed missing in *Remarks on the Foundations of Mathematics*; only that we must know where to look for it.

Much of the later part of the book is devoted to 'The idea of the mechanization of mathematics. The fashion of the axiomatic system' (RFM VII §12). Wittgenstein was not attacking the deductivist conception of mathematics, however, but rather was clarifying the logical grammar of 'One follows the rule *mechanically*. Hence one compares it with a mechanism. "Mechanical" — that means: without thinking. But *entirely* without thinking? Without *reflecting*' (RFM VII §60). This discussion has given rise to a heated controversy in the philosophy of language on the import and/or purport of Wittgenstein's remarks on the normative cluster of concepts which license the ascription of rule-following behaviour.[177] His argument has obvious importance for all those disciplines engaged in a 'cognitivist' rewriting of the logical grammar of rule-following.[178] At another level it bears fundamentally on the non-normative view of rule-following which inspired both Gödel and Turing's conception of algorithms.[179] But at the deepest level of all it suggests that given the rules of e.g. formal systems or Recursion Theory one has no choice but to accept the existence of true but unprovable propositions. The basic rules of arithmetic thus take on the appearance of railroad tracks which lead in one direction into the problem of hidden contradictions and in the other to Gödel's theorem. Or to take another of Wittgenstein's metaphors, once the machine has been set into motion neither Hilbert's sceptical dilemma nor Gödel's platonism can be avoided. To combat the larger problem of which these are but consequences Wittgenstein set out to remove the *Bedeutungskörper* conception of mathematical meaning;[180] for it is this picture which sustains Gödel's interpretation that 'a true but unprovable mathematical proposition' could be mechanically generated by the meaning of the primitive terms together with the transformation rules of a system.

The importance of this issue was borne out in an extremely interesting debate on the philosophical significance of the incompleteness theorems which, curiously, has been largely ignored by philosophers. In 1949 Irving Copi suggested that Gödel's theorem refutes the theory that all *a priori* truths are analytic. For 'if there is any non-empirical or non-inductive general proposition which is not decidable on the basis of the syntactical rules of the language in which it is expressed, then the analytic theory of *a priori* knowledge is false. ... And this amounts to saying that there are synthetic *a priori* propositions.'[181] Alarmed by this development Atwell Turquette promptly rejected Copi's attempt to impose such 'philosophical consequences' on Gödel theorem:

> The claim that there are Gödel synthetic *a priori* truths amounts to nothing more than a restatement in misleading philosophical language of some well-established logical

results, notably of what is usually called Gödel's second incompleteness theorem. The restatement in terms of a synthetic *a priori* truth is misleading since it suggests a connection with issues in the history of philosophy with which ... the Gödel theorems are very little concerned.'[182]

Copi subsequently clarified that 'The philosophical consequences do not flow from Wid, but *from the fact that there are such statements as Gödel's which are* a priori *true but not analytic.* The Gödel sentences were not our premises, but rather the fact that they have the properties which they have been proved to have.'[183] And there the debate ended.[184]

However much one might sympathize with Turquette's objectives it is clear that Copi had only clarified something which mathematicians — including and especially Gödel[185] — had long since recognized. The point which Copi successfully drove home is that it would be completely misguided to isolate philosophy from a mathematical demonstration that there are true but unprovable propositions. Certainly it would seem that the reason why Wittgenstein devoted so much attention to this issue was because of his anxiety that the latter might come to serve as a paradigm of synthetic *a priori* knowledge (RFM I App III §15). The Copi–Turquette debate is significant for an even more profound reason, however; for the heart of Turquette's argument was that

It is not unreasonable to believe that statements like Wid or ¬Wid are but instances of well formed formulas (grammatically correct formulas) which are nonsensical otherwise. Hence it may be more reasonable to assign no meaning to Gödel's undecidable statements than to assign to them the kind of meaning which is usually associated with synthetic *a priori* truth.[186]

To this crucial point Copi protested that 'we *do* understand the Gödel statements. We understand them on both interpretations which the arithmetization process provides us.'[187] This is not so much an argument as a reaffirmation of the standard interpretation; more important are the 'unhappy theoretical implications' which Copi drew from Turquette's objection:

The essential property of a language or symbolic *system* which distinguishes it from a mere collection of names is that the meanings of combinations of symbols are determined by the meanings of their constituent symbols and their mode of combination. The significant and valuable feature of a language is that one who knows its vocabulary and its syntax can understand the meanings of sentences which he has never encountered before. What Turquette's suggestion entails is that: *either* he must deny that any language in which Gödel undecidable formulas occur is a language in the sense explained above, *or* he must introduce a new type of formation rule to distinguish between well formed formulas which make sense and well formed formulas which are 'nonsensical'.[188]

Copi concluded that 'both horns of this dilemma are objectionable. To deny that Gödel and others are investigating language is to predicate irrelevance and triviality of the bulk of contemporary logical research.' But to demand 'a new type of formation rule of the type indicated' would make the sense of a mathematical proposition dependent on a proof of its truth.[189]

The dilemma as presented certainly cannot be lightly dismissed; the question is whether it is in fact generated by Turquette's objection or Copi's compositional theory of meaning. For one thing, the latter sits awkwardly on Hilbert's framework; as we saw in §5, Hilbert maintained that the meaning of mathematical propositions is assigned by interpretations: not built up out of the primitive rules of a system. But it fares far worse with Wittgenstein's willingness to take up both aspects of Copi's challenge. Philosophers of mathematics will need no reminder of Wittgenstein's infamous remarks on the manner in which '"Mathematical logic" has completely deformed the thinking of mathematicians and of philosophers (RFM V §48; cf. RFM V §46); while theorists of meaning are more than happy to ignore the sustained attack on the calculus conception of meaning that provides the foundation of *Philosophical Investigations*.[190] What is important here is how both sides of these major concerns of the later Wittgenstein came together in his remarks on Gödel's theorem. For the point on which Wittgenstein put most pressure is indeed — as Copi suggested must be the case — the generative picture of mathematical propositionhood which sustains the Gödelian response to the incompleteness theorems. On either of the two lines of interpretation outlined in the preceding section the ultimate question we are meant to ask is, as Wittgenstein stressed at RFM VII §22, what sort of *proposition* could $M_i$ be: i.e. how is it *used*, as opposed to how is it *constructed*.

The missing insight mourned by Goodstein is contained in the same passage which, significantly, is embedded in a prolonged discussion of the normativity of rule-following. The point is that even if the basic rules of a system result in the construction of a formula which, for whatever reason, we wish to place on the index — e.g. because it is or immediately leads to a contradiction[191] — that does not mean that this constitutes a case of the axioms forcing on us a mathematical proposition for which we not only do not but *could not* have any use. What we must 'remember [is] that in mathematics we are convinced of *grammatical* propositions; so the expression, the result, of our being convinced is that we *accept a rule*' (RFM III §26). It is not the rules of inference and the construction rules of a system which, on their own as it were, determine what shall count as a mathematical proposition; it is the use to which we do or can put such rules in the transformation of empirical propositions. The philosophical problem posed by Gödel's theorem thus comes down to the question: 'for what purpose do you write down this "assertion"? ... And how could you make the truth of the assertion plausible to me, since you can make no use of it except to do these bits of legerdemain?' (RFM I App III §19). The emphasis performed by the inverted commas here is crucial; for Gödel's theorem has 'the *form* of a proposition but we don't compare it with other propositions as a sign *saying* this or that, making *sense*' (RFM VII §22).

The contrast between *generation* and *deduction* operating here is deliberately tied in to the earlier distinction between logistic and mathematical proofs (see §5):

Here one needs to remember that the propositions of logic are so constructed as to have *no* application as *information* in practice. So it could very well be said that they were not *propositions* at all; and one's writing them down at all stands in need of justification. Now if we append to these 'propositions' a further sentence-like structure of another kind, then we are all the more in the dark about what kind of application this system of sign-combinations is supposed to have. (RFM I App III §20)

Seen in this light, Gödel's proof is a procedure for producing a formula which '*itself* is unusable': not for establishing the truth of a proposition. 'For the mere *ring of a sentence* is not enough to give these connexions of signs any meaning' (RFM I App III §20). But the fact that Gödel's proof might force the system to stall no more entails a sceptical than a metaphysical consequence: 'If it were like that, well, that is how it would be. (The superstitious dread and veneration by mathematicians in face of contradiction.)' (RFM I App III §17). This return to the running theme of the proposed resolution of the consistency problem both reminds us of the aetiology of Wittgenstein's remarks on Gödel's theorem and prepares us for the conclusion that 'Here once more we come back to the expression "the proof convinces us". And what interests us about conviction here is neither its expression by voice or gesture, nor yet the feeling of satisfaction or anything of that kind; but its ratification in the use of what is proved' (RFM VII 22).

As these last quotations make clear the present paper is but a prolegomenon to Wittgenstein's reflections on mathematical logic. The primary aim has been to locate Wittgenstein's remarks on Gödel's theorem within the larger context of his approach to the consistency and decision problems in order to show both how these obscure fragments fit into the overall strategy of his proposed resolution of the foundations crisis and thence the considerable background which they both presuppose and demand if they are to be seen to cohere. Perhaps the most important task to tackle next is to clarify Wittgenstein's attack on the notions of 'effective decidability' and 'mechanical calculation'. To have pursued these latter issues in any depth here, however, would have led dangerously into that area of exegesis where accusations of critical revisionism are most easy to prosecute. That is not to say that the materials for this discussion cannot be found in Wittgenstein's writings; only that we must look for them in a different quarter. But the problems and assumptions that belong to this subject open up another — interrelated — domain of concepts that demand meticulous and preferably non-polemical study. The insights contained in Wittgenstein's writings in the philosophy of mathematics have so far been barely touched upon. But in order for all concerned to benefit from his investigations it is imperative to learn from Wittgenstein's own example. For it was as a token of the high regard in which he held 'Gödel's greater service to the philosophy of mathematics' that Wittgenstein was so concerned to clarify the significance of the incompleteness theorems.

## NOTES AND REFERENCES

1. This paper discharges a self-imposed obligation from *Wittgenstein and the Turning-Point in the Philosophy of Mathematics* (London, Croom Helm, 1987). It should thus be read as a sequel to this earlier work, and as is hinted at in §7 below, a prolegomenon to the many important areas in mathematical logic where Wittgenstein's thought has yet to be clarified. It was originally Gordon Baker's idea to write this paper, but Michael Dummett should also be thanked for his encouragement. To Rand Hoenhous I owe a special debt for his invaluable assistance. It should be obvious how much I have benefitted from the authors gathered in this collection; special thanks are due to Michael Detlefsen, on whose writings and advice I have drawn throughout. Most important of all were the suggestions and criticisms which I received from Professor Angus MacIntyre, Rainer Born, Gordan Baker and Peter Hacker. As always my greatest debt is to my wife.

2. Ludwig Wittgenstein, *Remarks on the Foundations of Mathematics*, G.H. von Wright, R. Rhees, and G.E.M. Anscombe (eds.), G.E.M. Anscombe (trans.), 3rd edn (Oxford, Basil Blackwell, 1978).

3. As Harvard University described Gödel's theorem when awarding him an honourary degree in 1952. See Ernest Nagel and James R. Newman, *Gödel's Proof* (New York, New York University Press, 1958), p. 3.

4. James R. Newman, *The World of Mathematics*, Vol. 3 (New York, Simon and Schuster, 1956), p. 1616.

5. Nagel and Newman, *Gödel's Proof*, pp. 94-5.

6. Douglas R. Hofstadter, *Gödel, Escher, Bach: An Eternal Golden Braid* (Harmondsworth, Penguin Books, 1982); cf. I. Benardete, *Infinity* (Oxford, Clarendon Press, 1962) and R. Rucker, *Infinity and the Mind* (Brighton, Sussex, Harvester Press, 1982).

7. Currently placed at approximately 20,000 theorems per year.

8. David Hilbert, 'Mathematical Problems', trans. M.W. Newson, *Proceedings of Symposia in Pure Mathematics*, Volume 28 (1976), p. 1.

9. Ibid., p. 2. The uncharitable might feel that this is merely a muddled way of restating that the theorem should be significant.

10. And Poincaré; cf. Constance Reid, *Hilbert* (London, George Allen & Unwin, 1970), p. 68.

11. Ibid.

12. See *Mathematics of Computation*, Vol. XVI, no. 77 (January, 1962). The answer, I believe, is that there are four 7s in a row beginning at the 1,589th digit.

13. G.H. Hardy, *A Mathematician's Apology* (Cambridge, Cambridge University Press, 1969), pp. 119-20.

14. Ibid., p. 139.

15. A.N. Whitehead, 'Mathematics as an Element in the History of Thought', in Newman, *The World of Mathematics*, Vol. 1.

16. Hardy, *A Mathematician's Apology*, p. 89. Cf. Halmos:

> A good piece of mathematics is connected with much other mathematics, it is new without being silly ... and it is deep in an ineffable but inescapable sense — the sense in which Johann Sebastian is deep and Carl Philip Emmanuel is not. The criterion for quality is beauty, intricacy, neatness, elegance, satisfaction, appropriateness — all subjective, but all somehow mysteriously shared by all. (P.R. Halmos, 'Mathematics as a Creative Art', *Scientific American* (1958), p. 588)

17. Ludwig Wittgenstein, *Wittgenstein's Lectures on the Foundations of Mathematics*, Cora Diamond (ed.) (Hassocks, The Harvester Press, 1976), p. 47. ·

18. See Wantzel's paper on the 'Means of ascertaining whether a geometric problem can be solved with ruler and compasses' in *Liouville's Journal*, volume 2 (1837), pp. 366-72.

19. For an outline of Wantzel's proof, cf. Florian Cajori, 'Pierre Laurent Wantzel', in *Bulletin of the American Mathematical Society*, Vol. XXIV (1918), pp. 346-7.

20. Where Wantzel has been remembered is, in fact, precisely here. For the comparatively well-known 'Wantzel's Proof' of the impossibility of an algebraic solution for the general quintic equation refers to his 1845 paper 'De l'impossibilité de résoudre toutes les équations algébriques avec des radicaux', *Nouvelles Annales de Mathématiques*, Vol. 4 (1845), pp. 57-65.

21. Or perhaps penultimate, in so far as Galois theory was soon to provide what many regarded as the ultimate grounds for the impossibility of the trisection problem.

22. Quoted in E.T. Bell, *Men of Mathematics* (New York, Simon and Schuster, 1965), p. 311.

23. Ibid.

24. Ibid., p. 312.

25. A philosopher as temperamentally opposed to metaphysics as Waismann could still end up with a strikingly platonist position in the philosophy of mathematics, even though — or especially because! — this was not at all evident to Waismann himself; see S.G. Shanker, *Wittgenstein and the Turning Point in the Philosophy of Mathematics* chapter VIII.

26. It should, however, be noted that the examples have been chosen with an eye to avoiding constructivist issues as far as possible; not because the latter would undermine, but only because they would divert the present discussion. Many would no doubt argue that the Appel-Haken solution of the four-colour problem is also a case of a theorem with profound philosophical significance; but here the matter turns on the prior question of whether the Appel-Haken solution does,

in fact, yield a *theorem.* On both topics see *Wittgenstein and The Turning-Point in the Philosophy of Mathematics,* chapters II and IV.
  27. Rudolf Carnap, 'Autobiography', in *The Philosophy of Rudolf Carnap,* P.A. Schilpp (ed.), 1963, p. 53.
  28. Ibid.
  29. Ibid., p. 54.
  30. Carnap explained:

In the metalogic I emphasized the distinction between that language which is the object of the investigation, which I called the "object language", and the language in which the theory of the object language, in other words the metalogic, is formulated, which I called the "metalanguage". One of my aims was to make the metalanguage more precise so that an exact conceptual system for metalogic could be constructed in it. Whereas Hilbert intended his metamathematics only for the special purpose of proving the consistency of a mathematical system formulated in the object language, I aimed at the construction of a general theory of linguistic forms (Ibid.).

Cf. Karl Menger on the role of Gödel's theorem in the birth of metalogic in 'The New Logic', reprinted in *Selected Papers in Logic and Foundations, Didactics, Economics* (Dordrecht, D. Reidel Publishing Company, 1979), pp. 32-35.
  31. Köhler recounts that

the member of the Vienna Circle to whom Gödel expressed his platonism most pronouncedly was Rudolf Carnap, with whom Gödel had numerous conversations on 'metalogic' and foundations of mathematics around 1929-33. These discussions were centered around topics treated in *Logische Syntax der Sprache* (1934), which made extensive use of Gödel's incompleteness results. ... One of Carnap's principle aims was to avoid metaphysics inherent in object-theories of mathematics by translating all 'inhaltliche Redeweise', thus reducing Wittgenstein's logical form to physical structure of syntax. In their conversations, Gödel insisted that the 'inhaltliche Redeweise' was indeed admissible (e.g. when showing the unprovable formula of 1931 to be true), and that moreover the 'formale Redeweise' hence becomes unnecessary. Although Gödel had not persuaded Carnap on this fundamental issue, he *did* move Carnap in a strongly platonistic direction in his definition of analyticity, the capstone of the syntax program.

Eckehart Köhler, 'Gödel and the Vienna Circle: Platonism versus Formalism', in *History of Logic, Methodology and Philosophy of Science,* section 13 (Vienna Institute for Advanced Studies).
  32. See 'What is Cantor's Continuum Problem?', in P. Benacerraf and H. Putnam (eds.), *Philosophy of Mathematics: Selected Readings* (Englewood Cliffs, Prentice-Hall, 1964), p. 271.
  33. As Hilbert was to write to Frege in regards to the consistency

problem; see Gottlob Frege, *Philosophical and Mathematical Correspondence*, Gottfried Gabriel, Hans Hermes, Friedrich Kambartel, Christian Thiel, and Albert Veraart (eds.) (Oxford, Basil Blackwell, 1980), p. 42.

34. Cf. George Pitcher, 'Wittgenstein, Nonsense, and Lewis Carroll', in S.G. Shanker (ed.), *Ludwig Wittgenstein: Critical Assessments*, Vol. IV (London, Croom Helm, 1985).

35. Cf. Hofstadter, *Gödel, Escher, Bach*, p. 704.

36. Nor is Hofstadter alone. At the close of 'Consistency and Completeness — A Résumé' Frank DeSua concludes:

> In view of the lack of a universally acceptable proof, belief in the consistency of mathematics becomes then somewhat a matter of faith rather than reason. Suppose we loosely define a *religion* as any discipline whose foundations rest on an element of faith, irrespective of any element of reason which may be present. ... Mathematics would hold the unique position of being the only branch of theology possessing a rigorous demonstration of the fact that it should be so classified. (*American Mathematical Monthly*, 63 (1956), p. 305)

We shall return to this theme in the final section.

37. Hofstadter, *Gödel, Escher, Bach*, p. 254.

38. Cf. ibid., p. 251.

39. Cf. ibid., p. 255.

40. Viz. 'as a kind of intuitive conception which we can survey in its entirety in our mind's eye'. Michael Dummett, 'The Philosophical Significance of Gödel's Theorem', in *Truth and Other Enigmas* (London, Duckworth, 1978), p. 191.

41. Ibid., p. 186.

42. Ibid., p. 187.

43. See Shanker, *Wittgenstein and the Turning-Point in the Philosophy of Mathematics*, Chapter I.

44. Quoted in Hao Wang's *From Mathematics to Philosophy* (London, Routledge & Kegan Paul, 1974), pp. 8ff.

45. Ibid., p. 8.

46. Nagel and Newman, *Gödel's Proof*, p. 10.

47. Hilbert explained: 'The fundamental idea of my proof theory is none other than to describe the activity of our understanding, to make a protocol of the rules according to which our thinking actually proceeds.' (David Hilbert, 'On the Infinite', in *From Frege to Gödel: A Source Book in Mathematical Logic, 1879-1931*, Jean van Heijenoort (ed.) (Cambridge, Mass., Harvard University Press, 1967), p. 475.

48. Alan Ross Anderson, for example, complains:

> It is hard to avoid the conclusion that Wittgenstein failed to understand clearly the problems with which workers in the foundations have been concerned. Nor, I think, did he appreciate the real bearing of these results on the logistic thesis. For Gödel's theorem, and the line of development it culminates, have more often been cited in *favor* of Wittgenstein's position ('Mathematics and the

"Language Game"', in *Philosophy of Mathematics: Selected Readings*, Paul Benacerraf and Hilary Putnam (eds.) (Oxford, Basil Blackwell, 1964), p. 489).

49. Ludwig Wittgenstein, *Notebooks 1914-1916*, G.H. von Wright and G.E.M. Anscombe (eds.), G.E.M. Anscombe (trans.), 2nd ed. (Oxford, Basil Blackwell, 1979).

50. This is a very brief summary of a topic covered at some length in *Wittgenstein and the Turning-Point in the Philosophy of Mathematics*. It is primarily for reasons of space that readers are here referred to the discussion of Wittgenstein's proposed resolution of the Consistency Problem in chapter VI.

51. Marcus Giaquinto has challenged this interpretation, drawing on Gödel's cautionary note in 'On Formally Undecidable Propositions' that Theorem XI does not suffice to 'contradict Hilbert's formalist viewpoint' (see Chapter 2, this volume) to make the point that 'the intellectual pull of the belief in Hilbert's programme was, at least until 1931, so strong that even after his proofs of incompleteness and unprovability of consistency, Gödel himself felt disposed to defend the programme.' ('Hilbert's Philosophy of Mathematics', *British Journal of the Philosophy of Science*, vol. 34 (1983), p. 125). But Gödel's careful reading of the scope of Theorem XI should be read in conjunction with the 1931 'Postscript' which he prepared for *Erkenntnis*; in particular, his emphasis at the outset that 'On Formally Undecidable Propositions' 'is concerned with problems of two kinds, namely: 1. the question of the completeness (decisiveness) of formal systems of mathematics 2. the question of consistency proofs for such systems.' Gödel went on to stress that 'in all the well-known formal systems of mathematics — e.g., *Princ. Math.* ... the Zermelo-Fraenkel and von Neumann axiom systems for set theory, formal systems of Hilbert's school — there are undecidable arithmetical propositions.' (Quoted in John W. Dawson Jr., 'Discussion on the Foundation of Mathematics', *History and Philosophy of Logic*, vol. 5 (1984), p. 127.)

52. See *The Collected Papers of Gerhard Gentzen*, M.E. Szabo (ed.) (Amsterdam, North-Holland, 1969).

53. Hilbert, 'On the Infinite', p. 375.

54. Cf. Philip Kitcher, 'Hilbert's Epistemology', *Philosophy of Science*, vol. 43 (1976), pp. 99-115. In explaining Hilbert's motives for embarking on his Programme, Bernays described two 'tendencies in Hilbert's way of thinking. On one side, he was convinced of the soundness of existing mathematics; on the other side, he had — philosophically — a strong scepticism. The problem for Hilbert was to bring together these opposing tendencies, and he thought that he could do this through the method of formalizing mathematics.' (Quoted in Reid, *Hilbert*, pp. 173-4.)

55. It bears recalling how much prestige philosophy enjoyed at the time. It is interesting to note that when Hilbert read mathematics at the University of Königsberg it was still being taught as a sub-specialty in the philosophy faculty; indeed, Hilbert selected as one of the two theses which doctoral candidates were required to defend in public the propo-

sition 'That the objections to Kant's theory of the *a priori* nature of arithmetical judgments are unfounded'.

56. Quoted in Reid, *Hilbert*, p. 17.

57. First discovered by Bolzano; see Carl B. Boyer, *The History of the Calculus and its Conceptual Development* (New York, Dover Publications Inc., 1949), p. 270.

58. Nicht der Gödel'sche Beweis interessiert mich, sondern die Möglichkeiten, auf die Gödel durch seine Diskussion uns aufmerksam macht. Der Gödel'sche Beweis entwickelt eine Schwierigkeit, die auch in viel elementarerer Weise erscheinen muß. (Und hierin liegt, scheint es mir, (zugleich) Gödels großer Verdienst um die Philosophie der Mathematik, und zugleich der Grund, warum sein besonderer Beweis nicht das ist, was uns interessiert.

Quoted in *Wittgenstein: Sein Leben in Bildern und Texten*, Michael Nedo and Michele Ranchetti (Hrsg.) (Frankfurt am Main, Suhrkamp Verlag, 1983), p. 261.

59. Cf. Michael Resnik, 'On the Philosophical Significance of Consistency Proofs', chapter 6, this volume.

60. 'Ich könnte sagen: Der Gödel'sche Beweis gibt uns die Anregung dazu, die Perspektive zu ändern, aus der wir die Mathematik sehen. *Was* er beweist, geht uns nichts an, aber wir müssen uns mit dieser mathematischen *Beweisart* auseinandersetzen.' Quoted in Nedo and Ranchetti, *Wittgenstein: Sein Leben in Bildern und Texten*, p. 261.

61. 'Wenn Sie hören jemand haben bewiesen, es müsse unbeweisbare Sätze in der Mathematik geben, so ist daran *vorerst gar nichts Erstaunliches*, weil sie ja noch gar keine Ahnung haben, was dieser scheinbar so klare Prosasatz sagt. Sie haben also den Beweis von A bis Z. durchzugehen, um zu sehen, was er beweist.' Quoted in ibid., p. 260.

62. Quoted in Raymond Wilder's *Introduction to the Foundations of Mathematics* (New York, John Wiley & Sons, 1965), p. 265.

63. Ibid., p. 266.

64. Hardy, 'Mathematical Proof', p. 16.

65. At the beginning of 'Number and Magnitude' (*Science and Hypothesis*, New York, Dover Publications, 1952, pp. 1-2) Poincaré asked:

If ... all the propositions [of mathematics] may be derived in order by the rules of formal logic, how is it that mathematics is not reduced to a gigantic tautology? The syllogism can teach us nothing essentially new, and if everything must spring from the principle of identity, then everything should be capable of being reduced to that principle. Are we then to admit that the enunciations of all the theorems with which so many volumes are filled, are only indirect ways of saying that A is A?

This argument reduced the Logical Positivists to the expedient of arguing that to a being of infinite wisdom such would indeed be the case, and that the element of surprise involved in mathematical discovery was due to the finite limitations of human powers.

66. As with Wittgenstein's proposed resolution of the Consistency Problem (see Note 50 above) so here too, and for the same reasons, his proposed resolution of the Decision Problem is only touched upon; see *Wittgenstein and the Turning-Point in the Philosophy of Mathematics*, chapter III.

67. Hardy, 'Mathematical Proof', p. 16.

68. If nothing else, the publisher of such a text would be spared the cost of line-drawings.

69. See *Historisches Wörterbuch der Philosophie*, Joachim Ritter (Hrsg.) (Basel/Stuttgart, Schwabe & Co.), p. 1175.

70. See ibid.

71. Cf. Moritz Pasch, *Vorlesungen über neuere Geometrie* (Leipzig, B.G. Teubner, 1882), p. 98.

72. See David Hilbert, *Foundations of Geometry*, Leo Unger (trans.) (La Salle, Open Court, 1972), and the forthcoming paper by G.P. Baker and P.M.S. Hacker on 'Conventionalism'.

73. Cf. Thomas Heath, *A History of Greek Mathematics* (New York, Dover Publications Inc., 1981), pp. 235-44.

74. As was pointed out in the first section, had Wantzel's impossibility proof inspired Abel rather than the other way round how different would have been the former's subsequent renown.

75. Quoted in Reid, *Hilbert*, pp. 218-19.

76. G. Sarton, *The Study of the History of Mathematics* (New York, Dover, 1954), p. 17.

77. See Alan Ross Anderson, 'Mathematics and the "Language Game"', in Benacerraf and Putnam, *Readings in the Philosophy of Mathematics*, p. 487.

78. Wittgenstein was not entirely alone in the qualms which he felt about Hilbert's presentation of 'meta-mathematics'; for an unlikely kindred spirit, cf. Hans Hahn, 'The Scientific World View', in *Empiricism, Logic and Mathematics*, Brian McGuinness (ed.) (Dordrecht, D. Reidel Publishing Co., 1980), p. 26.

79. Hilbert, *Foundations of Geometry*, p. 2.

80. According to Roberto Torretti, although there were some efforts during the eighteenth-century 'to make explicit that which is tacitly understood in the *Elements*', 'no attempt at bringing out every presupposition of Euclid and filling all the gaps in his proofs was carried out in earnest until the end of the 19th century.' *Philosophy of Geometry from Riemann to Poincaré* (Dordrecht, D. Reidel Publishing Company, 1984), p. 189.

81. In the first of his recorded discussions with Waismann and Schlick Wittgenstein maintained:

> It must be borne in mind ... that it is impossible to move from one system to the other by merely extending the former; that a question which has sense in the second system need not therefore have sense in the first system. The new system is not a completion of the old one. The old system has no gaps. (WWK 36)

He then illustrated his point in terms of the unintelligibility of the elementary Euclidean version of the trisection problem.

82. Hermann Weyl, 'David Hilbert and his Mathematical Work', in Reid, *Hilbert*, p. 264.

83. Cf. Euclid, *The Elements*, Volume I, Thomas L. Heath (trans.) (New York, Dover Publications, 1956), pp. 241-2.

84. Ibid.

85. Cf. Weyl: 'If I am not mistaken, Hilbert is the first who moves freely on this higher "metageometric" level' (in Reid, *Hilbert*, p. 265).

86. But there is an enormous difference, Wittgenstein insisted, between saying that you have *demonstrated* this within the system and that you *proved* this in an entirely different system. Cf. *Philosophical Remarks*, p. 324.

87. R. Blanché, *Axiomatics*, G.B. Keene (trans) (London, Routledge & Kegan Paul, 1962).

88. In other words, we must distinguish between 'identical' and 'equivalent' systems. In the former the meaning of the totality of propositions remains the same and the only differences are ideographic; in the latter the only identity is homophonic. Jean Nicod, *Geometry and Induction* (Berkeley and Los Angeles, University of California Press, 1970), pp. 11-13. Max Black made a similar point in 'Conventionalism in Geometry and the Interpretation of Necessary Statements', *Philosophy of Science*, vol. 9 (1942), pp. 339ff.

89. Hilbert's letter to Frege of 29.12.1899, quoted in Frege, *Philosophical and Mathematical Correspondence*, p. 40.

90. An argument which in turn may owe far more to Nicod than has hitherto been recognized. A large part of *Geometry in the Sensible World* is devoted to elucidating the nature of 'global resemblance concepts' (pp. 59ff). Nicod's remarks on the manner in which 'Local resemblance acts as a kind of centre for a family of relations which cluster around it' (pp. 65ff) are especially worth attention.

91. The metaphor in this case is Bourbaki's, not Wittgenstein's; cf. 'The Architecture of Mathematics', *American Mathematical Monthly*, vol. 57 (1950).

92. Thus Torretti explains that 'The organization of geometry as a strictly deductive science, a collection of gapless axiomatic theories, was, of course, the natural way to deal with structures, because axiomatic theories are constitutively abstract.' *Philosophy of Geometry from Riemann to Poincaré*, p. 190.

93. David Hilbert, 'Mathematical Problems', in *Mathematics: People · Problems · Results*, volume I, Douglas M. Campbell & John C. Higgins (eds.) (Belmont, California, Wadsorth International, 1984), p. 277.

94. Ibid.

95. David Hilbert, *Foundations of Geometry*, p. 2.

96. As Mueller describes them in *Philosophy of Mathematics and Deductive Structure in Euclid's* Elements (Cambridge, Mass., The MIT Press, 1981), pp. 14 ff; cf. Chapter I passim.

97. Torretti, *Philosophy of Geometry from Riemann to Poincaré*, p. 229.

98. A premise which lent further credence to the belief that Hilbert had provided a 'foundation' for Euclidean geometry. Hilbert, *Foundations of Geometry*, p. 2.

99. Cf. Weierstrass: 'The final objective always to be kept in mind is to arrive at a correct understanding of the foundations.' Quoted in Reid, *Hilbert*, p. 82; cf. also the passage from 'The Future of Mathematics' quoted below, in which Poincaré makes very much the same point.

100. Frege, *Philosophical and Mathematical Correspondence*, p. 35.

101. Waismann, for one, interpreted the argument in this manner, but warned that it is 'certainly confusing to say that the primitive terms can be defined twice' ('The Structure of Concepts', in *Lectures on the Philosophy of Mathematics* (Amsterdam, Rodopi, 1982), p. 134. His own solution was to regard Hilbert's implicit definitions as a sort of 'pre-definition' which establishes the 'structure' of concepts, thus delimiting the range of all possible semantic interpretations. Hence 'There is something between "meaningless" and "having meaning in the full sense"' (p. 135). But far from resolving, this only reiterated Hilbert's problem.

102. Frege, *Philosophical and Mathematical Correspondence*, p. 42.

103. Ibid., p. 43. Although, as we shall in the following section, it is slightly misleading to suppose that Hilbert wanted to detach geometry entirely from spatial intuition.

104. Not, however, that as Judson Webb explains it in *Mechanism, Mentalism, and Metamathematics: An Essay on Finitism* (Dordrecht, D. Reidel Publishing Company, 1980), p. 81: 'any theory can only describe its basic domains up to isomorphism'. For the axioms of a system *describe* nothing; rather, they fix the meaning of the primitive concepts in the system. As Hilbert explained to Frege: 'a concept can be fixed logically only by its relations to other concepts. These relations, formulated in certain statements, I call axioms, thus arriving at the view that axioms ... are the definitions of the concepts.' (Quoted in Frege, *Philosophical and Mathematical Correspondence*, p. 51.)

105. Thereby generating the problem which preoccupied the Logical Positivists of accounting for the harmony between pure and applied geometry in terms which avoided the relapse into an empirical conception of geometry or a metaphysical conception of physical space; cf. S.G. Shanker, 'The Nature of Mathematical Reality', in *Essays in Honour of Q.A. Qadir*, N. Ahmed (ed.) (Lahore, University of Lahore), forthcoming.

106. Hilbert wrote to Frege, 'If the arbitrarily given axioms do not contradict one another, then they are true, and the things defined by the axioms exist. This for me is the criterion of truth and existence' (Frege, *Philosophical and Mathematical Correspondence*, p. 42).

107. By interpreting the 'points' and 'lines' of the system as pairs of real numbers and linear equations the consistency of the axioms could be demonstrated in the theory of real numbers.

108. Henri Poincaré, *Science and Method*, Francis Maitland (trans)

(London, Dover Publications, Inc.), p. 45.

109. Cf. Hilbert's late work, together with S. Cohn-Vossen, *Geometry and the Imagination*, P. Nemenyi (trans) (New York, Chelsea, 1952).

110. Quoted in Reid, *Hilbert*, p. 63.

111. See Henri Poincaré, 'Les mathématiques et la logique', *Revue de métaphysique et de morale*, vol. 13 (1905).

112. See David Hilbert, 'Neubegründung der Mathematik', in *Gessamelte Abhandlungen* (Berlin, Springer, 1935).

113. Hilbert complained:

Poincaré already made various statements that conflict with my views; above all, he denied from the outset the possibility of a consistency proof for the arithmetic axioms, maintaining that the consistency of the method of mathematical induction could never be proved except through the inductive method itself. But as my theory shows, two distinct methods that proceed recursively come into play when the foundations of arithmetic are established, namely, on the one hand, the intuitive construction of the integer as numeral ... that is, *contentual* induction, and, on the other hand, *formal* induction proper, which is based on the induction axiom and through which alone the mathematical variable can begin to play its role in the formal system.

'Poincaré', he sadly concluded, 'often exerted a one-sided influence on the younger generation' ('The Foundations of Mathematics', in van Heijenoort (ed.), *From Frege to Gödel*, pp. 471-2).

114. Cf. Jacques Herbrand, *Logical Writings*, Warren S. Goldfarb (ed.) (Cambridge, Mass., Harvard University Press, 1971), pp. 288-9.

115. Cf. Russell's review of *Science and Hypothesis*, *Mind*, vol. XIV (1905).

116. Cf. Henri Poincaré, 'On the Nature of Mathematical Reasoning', in *Science and Hypothesis* (New York, Dover Publications, 1952).

117. Cf. Webb, *Mechanism, Mentalism, and Metamathematics*, pp. 145f.

118. Cf. Hilbert, 'Die logische Grundlagen der Mathematik', p. 180, and 'On the Infinite', pp. 379ff.

119. It was for no idle reason that Hilbert was excited to learn that the laws governing the inheritance of characteristics in the fruit fly had been derived from an unrelated and pre-established set of axioms. Reid quotes Hilbert as having responded: 'So simple and precise and at the same time so miraculous that no daring fantasy could have imagined it!' *Hilbert*, p. 61. Here, he felt convinced, was proof of the validity of his answer to Poincaré.

120. Cf. E.T. Bell, *The Development of Mathematics* (New York, McGraw-Hill, 1945).

121. Nicholas Bourbaki, *Elements of Mathematics: Theory of Sets* (Paris, Hermann, Publishers in Arts and Science), p. 8.

122. Nicholas Bourbaki, 'The Architecture of Mathematics', *American Mathematical Monthly*, vol. 57 (1950), p. 222.

123. Ibid., p. 223.

124. Bourbaki, *Elements of Mathematics: Theory of Sets*, p. 7.

125. Hofstadter, *Gödel, Escher, Bach*, p. 35. Note the shift in this quotation from speaking of the 'mathematical' to the 'ordinary language' use of theorem: a subtle way of hinting that nothing more than a technical refinement is involved here. Note also the suggestion that formal theorems 'need not' — rather than *cannot* — be thought of as statements, thereby signifying that formal theorems are the shadows of and can thus be 'treated as' theorems.

126. 'A "free" theorem is called an *axiom* — the technical meaning again being quite different from the usual meaning.' Ibid.

127. See Shanker, *Wittgenstein and the Turning-Point in the Philosophy of Mathematics*, chapter VII.

128. This point is crucial when considering Wittgenstein's proposed resolution of the decision problem. For Wittgenstein did not deny the incompleteness of first-order functional calculus where to be complete (in the strong sense) entails that any wff is either a theorem or would render the system inconsistent if added to the axioms. Nor, conversely, did he intend his argument to be read as an endorsement of the effective decidability of all mathematical systems.

129. See Michael Detlefsen, *Hilbert's Program* (Dordrecht, D. Reidel, 1986), chapter one; cf. also the references listed in footnote 1 of the same chapter.

130. Hilbert, 'On the Infinite', p. 380.

131. Ibid., pp. 378, 380.

132. Cf. Detlefsen, *Hilbert's Program*, chapter 2.

133. Gottlob Frege, 'On the Foundations of Geometry', in E.-H. Kluge (ed.), *On the Foundations of Geometry and Formal Theories of Arithmetic* (New Haven, Conn., Yale University Press, 1971), p. 86.

134. Cf. Shanker, *Wittgenstein and the Turning-Point in the Philosophy of Mathematics*, chapter V.

135. Nor do any problems with the law of excluded middle result, for as Wittgenstein remarked in the context of his discussion of Skolem's proof, '$a + (b + c) = (a + b) + c \ldots A(c)$ can be construed as a basic rule of a system. As such, it can only be laid down, but not *asserted*, or denied (hence no law of the excluded middle)' (PR §163).

136. Hilbert, 'Neubegründung der Mathematik', p. 165.

137. Hilbert, 'On the Infinite', p. 383.

138. Wittgenstein's argument is thus strikingly reminiscent of Hilbert's letter to Frege (Note 89 above):

> The system of rules determining a calculus thereby determines the 'meaning' of its signs too. Put more strictly: The form and the rules of syntax are equivalent. So if I change the rules — seemingly supplement them, say — then I change the form, the meaning. (PR §152)

139. 'Let's remember that in mathematics, the signs themselves *do* mathematics, they don't describe it' (PR §157).

140. '*Frege* was right in objecting to the conception that the

numbers of arithmetic are signs, the sign '0', after all, does not have the property of yielding the sign '1' when it is added to the sign '1'. Frege was right in this criticism' (WWK 103).

141. 'Arithmetic is the grammar of numbers. Kinds of number can only be distinguished by the arithmetical rules relating to them. ... Arithmetic doesn't talk about numbers, it works with numbers' (PR §108-9; cf. PR §121).

142. Nagel and Newman, *Gödel's Proof*, pp. 28-9.

143. Ibid.

144. Ibid., p. 32.

145. Where a connection is now known to exist which was previously unknown, there wasn't a gap before, something incomplete which has now been filled in! — (At the time, we weren't in a position to say 'I know this much about the matter, from here on it's unknown to me.')

That is why I have said there are no gaps in mathematics. This contradicts the usual view. (PR §158)

146. See e.g. Nagel and Newman, *Gödel's Proof*, pp. 34ff.

147. See Shanker, *Wittgenstein and the Turning-Point in the Philosophy of Mathematics*, Chapter VI.

148. Nagel and Newman, *Gödel's Proof*, p. 78.

149. Cf. Wittgenstein's remark in *Philosophical Grammar*: 'Suppose that ... I say "love is a 2-place relation" — am I saying anything about love? Of course not. I am giving a rule for the use of the word "love" and I mean perhaps that we use *this* word in such and such a way' (PG, p. 310).

150. Nagel and Newman, *Gödel's Proof*, pp. 79-80.

151. Ibid.

152. Ibid., p. 82.

153. Kurt Gödel, 'On Formally Undecidable Propositions', chapter II above.

154. Nagel and Newman, *Gödel's Proof*, pp. 89-90.

155. This point might be considered in light of the debate in the philosophy of language whether syntax or semantics is responsible for the demarcation between *sense* and *nonsense*. For Wittgenstein this problem is the direct result of trying to divorce meaning from grammar. On his conception of logical syntax in *Philosophical Remarks* grammar determines the boundary between *meaningful* and *meaningless* constructions. Does the expression 'Green ideas sleep furiously' make no sense because it violates meaning-rules? The question is bizarre precisely because it does not make any sense to call this string of words a 'proposition'. For what exactly does it mean to assume that this string is syntactically well-formed but meaningless? 'You apply a calculus in such a way that it yields the *grammar* of a language. In grammar, then, the words "sense" and "senseless" correspond to what a rule permits and prohibits' (WWK 126). Cf. G.P. Baker and P.M.S. Hacker, *Wittgenstein: Understanding and Meaning* (Oxford, Basil Blackwell, 1980).

156. Cf. Anderson, 'Mathematics and the "Language Game"'.

157. As Gödel himself described Wittgenstein's remarks in a letter

to Abraham Robinson on July 2, 1973; see John W. Dawson, Jr., 'The Reception of Gödel's Incompleteness Theorems', chapter IV above.

158. Kurt Gödel, 'On Completeness and Consistency', in van Heijenoort, *From Frege to Gödel*, p. 617.

159. I. Grattan-Guiness, 'On the Development of Logics Between the Two World Wars', *American Mathematical Monthly*, Vol. 88, No. 7, p. 499.

160. Ibid. The comment from J. Barkley Rosser which Grattan-Guinness cites here is also extremely interesting, for according to Rosser, it was Gödel's theorem which, more than any other single work, sparked off the sudden widespread awareness and interest in meta-mathematics and recursion theory which quickly overtook mathematical logic.

161. No doubt some will feel that the argument moves too quickly here in so far as it equates platonism with metaphysics. Stephen Barker, to cite but one example, has argued in 'Realism as a Philosophy of Mathematics' that platonism need not, *qua* realist philosophy of mathematics, result in metaphysics. Barker suggests that realism is warranted on the grounds that it 'provides the simplest adequate explanation of why mathematics is intellectually significant'. Anyone tempted by such an argument should consider its similarity to the 'proofs' for the existence of phlogiston.

162. Cf. Post's remarks on 'mathematical creativity' in 'Recursively Enumerable Sets of Positive Integers and Their Decision Problems', in Martin Davis (ed.), *The Undecidable* (Hewlett, New York, Raven Press, 1965), pp. 315-6; cf. also Waismann's account of Gödel's theorem in *Introduction to Mathematical Thinking*, pp. 101f. Waismann's account is particularly interesting since, for reasons discussed in *Wittgenstein and the Turning-Point in the Philosophy of Mathematics*, it represents a conscious attempt to reconcile Wittgenstein's ideas with contemporary developments.

163. As in the case, e.g., of '5 + 7 = 12' and '(+)5 + (+)7 = (+)12'; cf. *Wittgenstein and the Turning-Point in the Philosophy of Mathematics*, Chapter III.

164. Charles Perelman, 'L'Antinomie de M. Gödel', *Academie Royale de Belgique, Bulletin de la Classe des Sciences*, Series 5, 22, p. 736.

165. Dawson, 'The Reception of Gödel's Incompleteness Theorems', p. 85.

166. Ibid., p. 87.

167. Cf. Olaf Helmer, 'Perelman versus Gödel', *Mind*, vol. 46 (1937), and Stephen Kleene's reviews of the latter and 'L'Antinomie de M. Gödel'.

The very fact that Perelman was unaware of the object/metamathematics distinction is more surprising than Dawson indicates, given the emphasis which Gödel placed on this crucial premise (in both the prose preface to 'On Formally Undecidable Propositions' and in 'On Completeness and Consistency') and the nature of the debate sparked off by Perelman's article. From a modern point of view this may seem an inexplicable oversight on Perelman's part but it provides a salutary reminder of the failure of Hilbert's 'meta-mathematical' revolution to make any immediate impact on mathematical and philosophical circles outside of Warsaw and Vienna. After all, it was by concentrating on the *mathematics* of Gödel's proof that Perelman was led to the conclusion that 'a formal study of the hypotheses which one finds at the base of Gödel's proof convinces us that Gödel's results are far more modest than might appear on first reading' (Ibid. p. 732). Presumably others were led into the same confusion by an instinct which might, perhaps, deserve more credit than hitherto accorded.

168. If only because he later went on to make his reputation in the virtually single-handed revival of Rhetoric.

169. John W. Dawson, Jr., 'Completing the Gödel-Zermelo Correspondence', *Historia Mathematica* 12 (1985), p. 69.

170. Cf. Ch. Perelman, 'L'Équivalence, la définition et la solution du paradoxe du Russell', *L'Enseignement Mathématique*, Nos. 5-6 (1937), pp. 35-6, 'Les Paradoxes de la logique', *Mind*, Vol. XLV, no. 178, pp. 204-8, and 'Une solution des paradoxes de la logique et ses conséquences pour la conception de l'infini', *Library of the Xth Congress of Philosophy*, Amsterdam (1938), pp. 206-10.

171. Although some might argue that he still did not apply it rigorously enough, Zermelo insisted in his lecture 'On the Levels of Quantification and the Logic of the Infinite' that 'a healthy "meta-mathematics," a true "logic of the infinite," first becomes possible through a *fundamental renunciation* of the assumption characterized above, which I term the *"finitistic prejudice".*' Quoted in G.H. Moore's 'Beyond First-order Logic: The Historical Interplay between Mathematical Logic and Axiomatic Set Theory', *History and Philosophy of Logic* I (1980), p. 126.

172. Ivor Grattan-Guinness, 'In Memoriam Kurt Gödel: His 1931 Correspondence with Zermelo on his Incompletability Theorem', *Historia Mathematica*, vol. 6 (1979), p. 301.

173. Ibid., p. 302.

174. See Dawson, 'The Reception of Gödel's Incompleteness Theorems' and Feferman, 'Kurt Gödel: Conviction and Caution'.

175. Albert William Levi, 'Wittgenstein as Dialectician', *The Journal of Philosophy*, vol. LXI, no. 4 (1964), p. 134.

176. R.L. Goodstein, 'Critical Notice: *Remarks on the Foundations of Mathematics*', *Mind*, vol. 66 (1957), p. 551; cf. Goodstein's 'The Significance of Incompleteness Theorems', *Bulletin of the Journal of the Philosophy of Science*, vol. 14 (19), pp. 210ff.

177. See G.P. Baker and P.M.S. Hacker, *Scepticism, Rules and Language* (Oxford, Basil Blackwell, 1984).

178. See S.G. Shanker, 'The Decline and Fall of the Mechanist Metaphor', in Rainer Born (ed.), *AI: The Case Against* (London, Croom Helm, 1987), 'Computer Vision or Mechanist Myopia?', in S.G. Shanker (ed.), *Philosophy in Britain Today* (London, Croom Helm, 1986), and 'AI at the Crossroads', in Brian Bloomfield (ed.), *Questions in AI* (London, Croom Helm, 1987).

179. See S.G. Shanker, 'The Nature of Philosophy', in S.G. Shanker (ed.), *Ludwig Wittgenstein: Critical Assessments*, vol. IV (London,

Croom Helm, 1985).

180. Cf. Shanker, *Wittgenstein and the Turning-Point in the Philosophy of Mathematics*, chapter VII.

181. Irving M. Copi, 'Modern Logic and the Synthetic *A Priori*', *The Journal of Philosophy*, vol. XLVI (1949), p. 244.

182. Atwell R. Turquette, 'Gödel and the Synthetic *A Priori*', *Journal of Philosophy*, Vol. XLVII (1950), p. 125.

183. Irving M. Copi, 'Gödel and the Synthetic *A Priori*: A Rejoinder', *Journal of Philosophy*, vol. XLVII (1950), p. 634.

184. Apart from the noteworthy contribution by Jan Wolenski, 'Metamathematics and Philosophy', *Bulletin of the Section of Logic*, vol. 12 (1983).

185. Cf. Kurt Gödel, 'Russell's Mathematical Logic', in Benacerraf and Putnam, *Philosophy of Mathematics*, 2nd edition, pp. 449f.

186. Turquette, 'Gödel and the Synthetic *A Priori*', p. 129. This is a problem which has indeed troubled philosophers of mathematics; cf. L. Goddard, '"True" and "Provable"', *Mind*, vol. 67 (1958), Charles Parsons and Herbert R. Kohl, 'Self-Reference, Truth and Provability', *Mind*, vol. 69 (1960) and Richard Butrick, 'The Gödel Formula: Some Reservations', *Mind*, vol. 74 (1965).

187. Irving M. Copi, 'Gödel and the Synthetic *A Priori*: A Rejoinder', *Mind*, vol. XLVII (1950), p. 635.

188. Ibid.

189. Ibid., pp. 635-6.

190. See G.P. Baker and P.M.S. Hacker, *Wittgenstein: Understanding and Meaning* (Oxford, Basil Blackwell, 1980), and *Language, Sense and Nonsense* (Oxford, Basil Blackwell, 1984).

191. To reiterate the conclusion reached in §6, depending on which route of Wittgenstein's critique we take $M_i$ will be seen as either a significant (well-formed) but trivial proposition which is mapped onto a 'corresponding' nonsensical (ill-formed) expression (U), or else a senseless pseudo-proposition in which all information is cancelled out.

192. I make a start on these issues in 'Wittgenstein versus Turing on the Nature of Church's Thesis', *Notre Dame Journal of Formal Logic* (forthcoming).

# Index